**Stripe
Press**

Ideas for progress
San Francisco, California
press.stripe.com

Scientific Freedom:
The Elixir of Civilization

For Thomas Edward, and Christopher Jack.

.

Also by Donald W. Braben:

To Be a Scientist: The Spirit of Adventure in Science and Technology

Pioneering Research: A Risk Worth Taking

Promoting the Planck Club: How Defiant Youth, Irreverent Researchers and Liberated Universities Can Foster Prosperity Indefinitely

Published in the United States of America
by Stripe Press / Stripe Matter Inc.

Previously published in paperback

Stripe Press
Ideas for progress
San Francisco, California
press.stripe.com

Printed by Hemlock in Canada
ISBN: 978-0-578-67591-6

The printing of this book is carbon neutral in
partnership with Hemlock Printers and Offsetters.

Second Edition

Table of Contents

List of Posters

12

Introduction to New Edition

Since the Industrial Revolution, and especially since the beginning of the twentieth century, the idea of the inevitability of scientific progress has been powerful. Scientists could be relied upon to provide a steady trickle of unexpected major discoveries that would transform lives for the better. However, the past few decades have seen a major change. Nowadays, the funding of academic research is controlled by the consensus of experts, and peer review is by far the most common process for reaching consensus.[1] The progress of science, therefore, depends on the support that researchers can find.

The great iconoclastic scientist Albert Szent-Györgyi was the first to isolate the vitamin C molecule and to discover the components and reactions of the citric acid cycle, for which he won the Nobel Prize in 1937. In a letter to the journal *Science* in 1972, he said that scientists could be divided into two classes, Dionysians and Apollonians—in science, the Apollonian tends to develop established lines to their limit, while the Dionysian relies on intuition and is more likely to open new, unexpected lines of research.[2] Szent-Györgyi, of course, regarded himself as a Dionysian.

The funding agencies in every country in the world, but especially Britain, seem to believe that a healthy scientific enterprise can best be arranged if agencies rigorously concentrate funding on those well-defined areas deemed the highest priorities, thus strongly favoring Apollonians. Consequently, academic research is becoming increasingly predictable, which encourages competitiveness and focuses on the short term. If funding is to be severely constrained, as it is today, decisions on whom to fund should be based on the quality of proposals, as judged by the most absolute standards that can be arranged. This strategy would ensure that generating understanding extends over the widest possible spectrum, and would

---

1 I have attempted to take a global approach to this problem. There is a prodigious literature, of course, and such terms as "peer review," "merit review," and "expert review" are used interchangeably. The literature mainly focuses on the methods that might improve their efficiency. See, for example:

Sandra Bendiscioli, "The Troubles with Peer Review for Allocating Research Funding," *EMBO Reports* 20, no. 12 (2019), https://doi.org/10.15252/embr.201949472.

The Organisation for Economic Co-operation and Development (OECD), "Enhancing Research Performance through Evaluation, Impact Assessment and Priority Setting," 2008, https://www.oecd.org/sti/inno/Enhancing-Public-Research-Performance.pdf.

2 Albert Szent-Györgyi, "Dionysians and Apollonians" letter, *Science* 176, no. 4038 (1972): 966, https://science.sciencemag.org/content/176/4038/966.1

be more tolerant of the dissent so essential to progress. Instead, decisions on whether to fund proposals are determined by the pronouncements of a few fellow scientists, who are often applicants' closest competitors. Applicants are also frequently required to assess the future socioeconomic benefits of their proposed research and the steps they will take to realize them, which further constrains prospects for the unpredictable.

Funding success rates are in the region of 20%, taking universities' own arrangements into account. Many excellent proposals are lost, and funding agencies have strict rules on resubmission. On other days, with other experts with different perspectives making the fateful decisions, those might have been funded.

The universe is apparently infinitely complex, and our understanding of it might not always be reached by logical progression. We understand only a tiny proportion of what there is to be understood, and some discoveries have to be stumbled upon by scientists who must also recognize their luck. The best way of arranging for that, as we have seen over the past century and more, is to allow the most creative scientists freedom to roam, explore, and follow their intuition. Attempts to centrally manage these processes will cause them to fail. These processes cannot be managed by objectives.

Facing rejection rates of around 80% and having spent many anxious months preparing their proposals, scientists are likely to avoid the so-far intractable problems, because progress might require radically new approaches, assumptions, or techniques that will probably not fit into the current priorities or otherwise not command colleagues' full support. However, when unfashionably audacious proposals eventually turn out to be successful, as they often have been, they can open up new fields that lead to new insights and—for agencies interested in gainful outcomes—opportunities for highly profitable investment.

How has this situation come about? The answer can be found in the global quest since about the 1970s to considerably expand both the number of universities and the number of people attending them.[3] These changes have been made largely for political reasons—to sharply increase the nation's share of graduates, for example. As a consequence, universities have become less elitist—but scientists who would seriously challenge the foundations of knowledge will, if successful, inevitably become regarded as elite, which, of course, will work against this policy. The rapidly increasing number of academics has also made it necessary for funding

---

3   Evan Schofer and John W. Meyer, "The Worldwide Expansion of Higher Education in the Twentieth Century," *American Sociological Review* 70, no. 6 (December 2005): 898–920, https://www.jstor.org/stable/4145399?seq=1.

agencies to introduce stringent measures to ration research support. Research *proposals* are now treated as if they were completed works, and the so-called best, as selected by peer review, determine the research that gets done. This inhibits the scope of scientific progress.

If in the early to mid-twentieth century research funders had used similar policies to those prevailing today, they might have sought such surprise-free objectives as more efficient thermionic valves, aircraft piston engines, or new uses for coal, using operational research as a guide. Thankfully, inspired by such visionaries as Vannevar Bush in the US and Henry Dale in the UK, researchers remained largely free, and, provided the resources required were modest, they did not have to submit their plans to third parties: they simply went ahead with them. Many successfully tackled the so-called fashionable problems, but this freedom also enabled about 500 radical researchers to confront some unsolved problems. They were mostly academics and went on to win Nobel Prizes, an august gathering I have called the Planck Club, in honor of Max Planck, who discovered quantization in 1900.

Scientific opinion was important even then. Planck once said that science advances one funeral at a time: "A new scientific truth does not triumph by convincing its opponents and making them see the light, but rather because its opponents eventually die, and a new generation grows up that is familiar with it."[4] Hence the folly of using peer review to assess virtually all proposals, the current custom. The role of young researchers—their exuberance, confidence, and tendency to ignore tradition—is as vitally important today as it always has been. Many of the great discoveries were made by scientists in their twenties or thirties,[5] but for such youth to get unconstrained backing would be almost impossible nowadays. Supporting data is not easy to obtain, but the age at which researchers funded by the National Institutes of Health, the United States' largest research funder, receive their first grant rose from 38 years in 1980 to over 45 in 2013.[6] The situation seems much the same elsewhere and in other disciplines. This trend must surely be reversed.

Most Planck Club discoveries were unforeseen and were often not accepted by peers even after they had been demonstrated. Ernest Rutherford, who together with his eventually famous collaborators discovered

---

4   Max Planck, *Scientific Autobiography and Other Papers* (New York: Philosophical Library, 1950), 33.
5   Table A lists some discoveries made by young people. The # symbol denotes a scientist's loss of four years due to the Great War, 1914–1918.
6   R. J. Daniels, "A Generation at Risk: Young Investigators and the Future of the Biomedical Workforce," *Proceedings of the National Academy of Sciences* 112, no. 2 (January 2015): 313–318, https://doi.org/10.1073/pnas.1418761112.
   See also Figures 4 and 5 in *Scientific Freedom*.

ways of extracting power from nuclear reactions, declared in 1933 that anyone who thought that this feeble energy source would lead to practicable applications was talking moonshine;[7] Max Perutz and John Kendrew's discovery in the 1950s of the structures of hemoglobin and myoglobin,[8] protracted because of its difficulty over decades, took place in a *physics* laboratory and laid the foundations of protein crystallography, which played a vital role in biotechnology and the pharmaceutical industry; Peter Mitchell's discovery in the 1960s of chemiosmosis,[9] one of the most important advances in twentieth-century biochemistry, was bitterly opposed by almost everyone for many years; and there were numerous new medical diagnostics, such as magnetic resonance imaging, discovered in the 1970s by Peter Mansfield[10] and Paul Lauterbur.[11] Their ideas had not been included in the mainstreams but their respective universities (Nottingham and Stony Brook) gave them freedom to explore.

These discoveries are all in the natural sciences, but for many years academics have tried to explain the causes of growth from an economic point of view without reference to specific discoveries: since the 1950s, one of the most widely accepted concepts has been "creative destruction," derived by Joseph Schumpeter,[12] often called Schumpeter's gale. This concept postulates an incessant struggle among industries in which outdated old technologies are replaced by selections from the new, the threat of which stimulates industrial innovation. Although there is no general agreement among scientists and economists on the causes of growth, it may be said that in today's world, with industry's rapidly growing influence on academic research and the latter's now total dependence on peer review and consensus (which are not conducive to the creative processes), the outlook for unexpected research-led growth would seem bleak.

Planck Club discoveries could justifiably be called heroic because they flew in the face of accepted opinion. They also, apparently, gave enormous boosts to economic growth, and explain why we are well placed today. Some $100 trillion in today's money seems a conservative estimate of the twentieth-century global value of these discoveries, but they also

---

7   *The Times* archives, "The British association—breaking down the atom," September 12, 1933.

8   J. C. Kendrew, G. Bodo, H. M. Dintzis, R. G. Parrish, H. Wyckoff, and D. C. Phillips, "A Three-Dimensional Model of the Myoglobin Molecule Obtained by X-Ray Analysis," *Nature* 181 (March 1958): 662–666, https://doi.org/10.1038/181662a0.

9   Peter Mitchell, "Coupling of Phosphorylation to Electron and Hydrogen Transfer by a Chemi-Osmotic Type of Mechanism," *Nature* 191 (July 1961): 144–148, https://doi.org/10.1038/191144a0.

10 P. K. Grannell and P. Mansfield, "Microscopy in Vivo by Nuclear Magnetic Resonance," *Physics in Medicine and Biology* 20, no. 3 (1975): 477-82.

11 P. C. Lauterbur, "Image Formation by Induced Local Interactions: Examples Employing Nuclear Magnetic Resonance," *Nature* 242 (1973): 190–191, https://doi.org/10.1038/242190a0.

12 Hugo Reinert and Eric S. Reinert, "Creative Destruction in Economics: Nietzsche, Sombart, Schumpeter," in *The European Heritage in Economics and the Social Sciences* Vol 3, ed. J. G. Backhaus and W. Drechsler (Boston: Springer, 2006), 55–85.

contributed vast intangible benefits to quality of life. Under today's rigid rules, many of these visionaries would probably not be funded when they were setting out. Their twenty-first-century would-be successors will therefore face immense difficulties.

Most funders seem aware of this serious problem and have developed policies that they believe are more risk tolerant.[13] However, they seem unable to abandon faith in peer review or to accept viable alternatives. While peer reviewers are invited by the committees that oversee them to be more adventurous and encourage proposals from the young, peer review remains stubbornly conservative and inhibitory of radical initiative. Today, peer review of research proposals has been accorded gold-standard status—a benchmark of the highest quality—which places it, in effect, beyond question. Its endorsement is also thought to be the only way of ensuring research excellence. These major changes have therefore passed into acceptance without real debate. In the UK, in Europe, and elsewhere, academics acquiesce simply because there is little alternative. Largely monolithic funding arrangements do not help. The US, on the other hand, has a wide range of funding sources, but America's major grant-making bodies still rely almost totally on peer review, as do many other countries.

What should be done? It is proposed that potential pioneers—possible members of a twenty-first-century Planck Club—be selected and funded by individual universities from their own resources, as indeed used to be the practice. Though not discussed in *Scientific Freedom*,[14] my proposal focuses attention on those very rare staff whose research would radically challenge what we think we know. Each participating university should appoint one or two scientists to make the selections, avoiding, of course, the influence of peer review, consensus, and impact. They should be senior and mostly will have withdrawn from active research. Radically new policies will be needed for these initiatives, which will not be easily achieved. Approval rates will be very low, not for lack of funds but because the standards set will be exceptionally high. Most years, most universities will probably not find an applicant who satisfies the exacting standards, but the processes should be seen as exciting—in effect, universities will be attempting to select potential future Nobel Prize winners (or their equivalent)! There will be mistakes and misjudgments, but at least some radical

---

13 David Willetts, *The Road to 2.4 Per Cent: Transforming Britain's R&D Performance* (King's College London: The Policy Institute, December 2019). Willetts claims that the absence of peer review would make it easier to stick to a strategy, such as antimicrobial resistance, but seems to ignore the fact that sticking to a strategy is part of the problem posed by peer review. Willetts was the UK's minister of state for universities and science from 2010 until July 2014.

14 The proposal is discussed in my book *Promoting the Planck Club: How Defiant Youth, Irreverent Researchers and Liberated Universities Can Foster Prosperity Indefinitely*, Wiley 2014.

applications will survive, and some of them will succeed. Selectors should seek to redefine the role of research, to back imaginative people not projects, and, above all, to listen.

Total freedom should be given to some pioneers, but how can we reliably identify those who should have it? There are many scientists and many fields. One does not even know which haystack hides the needle. It's been forgotten that we did not need special arrangements for finding the Einsteins in the past. There was enough flexibility in the system to allow them to emerge, but that's been removed in the quest for efficiency.

Selectors might seek, for example, to reproduce the conditions enjoyed by members of the twentieth-century Planck Club when they were setting out—except, of course, that freedom for all now seems impossible. Universities should invite applicants to submit one-page précis of their new concepts, what they want to do, why they want to do it, and why they might not be funded by the usual agencies. Alternatively, applicants could be invited to discuss their ideas informally with a selector, thus avoiding all bureaucracy. There would be no deadlines. Selectors should be prepared to listen critically but sympathetically in extended, face-to-face discussions with applicants, and offer feedback in real time. They should aspire to standards independent of fashion, and should attempt to forge close relationships with applicants, whether or not these persons might be successful at gaining funding. For the sprinkling of proposals approved, funding's red tape would be considerably reduced, and the process would provide a viable alternative to current definitions of research excellence. For the selectors, they should take vicarious pleasure in the discoveries of others. The scientists they select will be extraordinary people, but, nevertheless, their proposals would almost certainly not attract funding from the usual sources.

The selection techniques outlined here, based on British Petroleum's successful Venture Research initiative, which operated from 1980 to 1990, are fully discussed in *Scientific Freedom*. An initiative that cost BP less than £20 million ($40 million) over the decade has yielded an estimated value approaching $1.25 billion today. We selected some 40 proposals over the course of 10 years (from approximately 10,000 applicants), of which at least 14 radically changed the way we think on an important topic. Unfortunately, however, this did not stop a newly appointed BP managing director from closing down the program in 1990, on the grounds that BP could no longer support the "drain on its resources"! Some of the initiative's many successes could be foreseen even then, a few years before the commercial profitability of the ideas it spawned became a demonstrable fact. They included Steve Davies, at the University of Oxford, whose Venture

Research project (1985–91) "Understanding Molecular Architecture" led to the development of small artificial enzymes for efficient chiral selection. This led him in 1992 to set up a company, Oxford Asymmetry, which he sold in 2000 for £316 million ($475 million) to a German firm, Evotec. He has since gone on to found six companies in other areas of science while remaining an academic committed to research. Other successes and honors won by Venture Researchers are given in the references to this Introduction.[15]

Costs of new initiatives (usually referred to in *Scientific Freedom* as transformative research rather than Venture Research, following the US lead) are also discussed, but after many years spent unsuccessfully searching for industrial or private sponsors for such initiatives, I have since 2008 changed emphasis to individual universities acting independently, the reason why the present proposal is not discussed in *Scientific Freedom*. Science does not lack opportunity. There are few, if any, fields that are fully understood. Thus, radical approaches on, say, the nature of gravity, or the sources of dark energy/dark matter, or consciousness, or any topic whose importance has yet to be widely recognized might qualify for support. Scientific potential, therefore, is almost as great as it was, say, a hundred years ago. But we shall create a twenty-first-century Planck Club and its spectacular harvest of unforeseen breakthroughs only if we restore the freedom that stimulates them.

A similar scheme currently operates for staff at University College London (UCL), funded from its own resources. Set up in 2008, the program has received some fifty proposals, from which so far one has been selected—that of Nick Lane for a largely theoretical study of the role of mitochondria in cells, costing the equivalent of about $275,000 over three years. It has since been considerably expanded to study the origin of life, and its growing potential has attracted the equivalent of around $2.5 million in external funding, around 10 times UCL's initial outlay. Lane has now joined UCL's professorial staff, created new approaches to this unsolved problem, and attracted similarly uninhibited young scientists from all over the world.

Unfortunately, university finances today are under unprecedented pressure, and many might find it difficult to embark on new programs, no matter how inexpensive they may prove. On the plus side, original thinkers—Venture Researchers—tend to be inspirational because they are tackling difficult problems they have chosen, perhaps using novel techniques, and like many of their Planck Club predecessors are paying scant regard to

---

15 See Tables B and C for successes and honors won by Venture Researchers.

where their inquiries may take them. Contrast that with what tends to be the norm these days: approximately 20% of proposals are successful, and their authors are increasingly encouraged to build bigger networks that might better compete, to attempt to anticipate every problem that might arise as the research progresses. These undertakings are decreed the best value for money, and probably will lead to only incrementally beneficial outcomes.

It is possible there may be another explanation. It is almost impossible to gain acceptance for Venture Research—the idea of unfettered research with radical objectives performed as a distinct and separately funded initiative—because it seems deeply believed to be unnecessary. Funding agencies, journals, and institutions apparently firmly hold that current arrangements adequately cater for every such departure. Peer review totally dominates thought, and it seems inconceivable that it could sometimes be seriously wrong. However, Venture Research proposals lie beyond the mainstreams of research so that they do not yet have peers: their proposals are unique and deserve special treatment. UCL's initiative was nevertheless brought about only after a long campaign and strong and sustained support from David Price, the vice provost for research, and other senior staff.

Universities would act independently. In the future, there might be merit in their coming together to compare selection techniques in this difficult area, offering a chance for researchers to share their crusades, just as was done in the BP-funded initiative to great mutual benefit. Real-time ancillary benefits, such as improvements in graduate recruitment and positive media responses, more than repaid its costs. Such research reveals its inherent potential and provides ample justification for the special measures needed for its selection.

Although universities should take the lead, as almost all the major Planck Club discoveries originated in the academic sector, industries (and particularly large companies such as BP, IBM, and Roche) and private investors and philanthropists should reconsider their sponsorship of radical basic research conducted by individuals or small groups. In today's world, which requires researchers to justify their support in terms of its possible impact, such lonely mavericks struggling in the dark recesses of ignorance need that vital help and the encouragement it provides. The costs would be relatively small, as this type of research is remarkably cheap, but, as with the universities, there would be considerable policy changes needed to allow them to sponsor research that offers the prospect of unpredictable outcomes from the start. As ever, well-connected champions for these initiatives will be essential.

Venture Research, therefore, has huge potential and low cost, but those selecting the research will have the same uncertainties and doubts as those faced by the selected researchers themselves. In today's world, the fashionable tendency is to see value in cost and to dismiss low-cost programs as unimportant or irrelevant, which would be to ignore the BP and UCL experience. If the selected research does not show signs of having high potential after, say, five years, its support should be terminated. However, it is vital that Venture Research programs are protected against emergent changes in policies, so that selectors can maintain the necessary enthusiasm and commitment.

In my long experience since leaving BP, I have found that the near reverence for peer review and established thinking in general is virtually confined to the policy-making ranks in the organizations that support research, and to the governments and private funding agencies they have persuaded to accept the infallibility of peer review. By focusing on such externalities as efficiency and value for money, these organizations favor Apollonians and seriously limit the creativity of the proposals they will support. At the level of the working scientist, these policies are treated as facts of modern life that must be tolerated if one wants funding. Unfortunately, that tolerance apparently morphs into acceptance when scientists are "promoted" into these ranks, which may explain these organizations' unchanging behavior over the years, and why they seem insulated against radical change.

In a little over half a century, therefore, the administration of science has been transformed, as I explain in Chapter 4. The role of the center, at least in the UK, which once strongly advocated policies Bush and Dale would have endorsed, has mutated into an indicator of the policies and objectives deemed to be in the national interest. Little else gets support.

Today, many of humanity's pressing problems—such as climate change, cyber threats, demographic change, economic growth, energy generation, income inequality, pollution, and the ever-present dangers of global conflict—have significant scientific components. Understanding is essential. Problems understood are much easier to solve.

Chapter 1 outlines some of the properties of the universe we inhabit, the most important of which seems to be its intrinsic nonlinearities. It is possible that we may be seeing early signs in parts of the vast continent of Australia that their effects may be pushing us toward the terrifying Damocles Zone. However, the consequences of nonlinearities are not only unforeseeable: they may flare up elsewhere in time and place.

Politics and fashion dictate most things, but there is an urgent need for the *unpredictable* benefits and insights that unconstrained science can bring. This modest proposal, by which universities are invited to follow UCL's lead, is one effective and consensus-free step toward achieving that end. Its modest cost should, in principle, favor its selection, but, as we have noted, that may not always be seen as an advantage.

As I write, we are being subjected to waves of restrictions designed to minimize the effects of SARS-CoV-2 and COVID-19, the names given to the virus and disease that threaten humanity. Let's hope the restrictions succeed. When the immediate threat recedes, there will be an even greater need for Venture Researchers. Their work has been proved to expand horizons and inspire investors to create the opportunities that lead to the additional growth the world desperately requires.

I would like to thank Nick Lane for his close friendship, as well as David Price and many friends in UCL's Earth Sciences Department. I would also like to thank John Allen, Terry Clark, Merlin Crossley, John Dainton, Steve Davies, Rod Dowler, Felipe Fernandez-Armesto, Desmond Fitzgerald, Pat Heslop-Harrison, Dudley Herschbach, Sui Huang, James Ladyman, Peter Lawrence, Chris Leaver, John Mattick, Jeff Kimble, Douglas Randall, Rich Roberts, Ken Seddon, Bill Troy, Robin Tucker, Colin Self, Tim Spiller, Harry Swinney, and Claudio Vita-Finzi for their sustained support over the years. The loving devotion of my wife, Shirley, and the support of our children and their spouses, David and Wendy, Peter and Lisa, and Jenny and David, have played a vital role in preserving my sanity in the face of the establishment's implacable opposition.

Donald W. Braben
d.braben@ucl.ac.uk
April 2020

List of Acronyms

CBE	Commander of the Most Excellent Order of the British Empire
DBE	Dame Commander of the Most Excellent Order of the British Empire
FREng	Fellow of the Royal Academy of Engineering
FRS	Fellow of the Royal Society
KB	Knight Bachelor
KBE	Knight Commander of the Most Excellent Order of the British Empire
NAS	US National Academy of Sciences
NAE	US National Academy of Engineering
OBE	Officer of the Most Excellent Order of the British Empire

Table A: Fruits of Exuberant Youth

Name, Year of Discovery	Discovery	Age at Discovery
Max Planck, 1900	Energy quantization	42
Albert Einstein, 1905	Photoelectric effect	26
Frederick Soddy, 1912	Isotopes.............................	35
Niels Bohr, 1913................	Hydrogen spectrum	28
Satyendra Nath Bose, 1923	Bose-Einstein statistics, Bosons	29
Louis de Broglie, 1924	Wave-particle duality	32 #
Wolfgang Pauli, 1925	Exclusion principle	25
George Uhlenbeck and Samuel Goudsmit, 1925	Electron spin	25, 23
Enrico Fermi, 1926.............	Fermi-Dirac statistics, Fermions	25
Erwin Schrödinger, 1926	Electron wave equation	39 #
Werner Heisenberg, 1927	Uncertainty principle	26
Paul Dirac, 1928	Atomic theory: predicted positron	26
James Chadwick, 1932	The neutron	41 #
Linus Pauling, 1932	The nature of chemical bond	31

The # symbol denotes a scientist's loss of four years due to the Great War, 1914–1918.

Table B: Venture Research Successes

Name	Success
Mike Bennett and Pat Heslop-Harrison, Plant Breeding Institute, 1986–92	Discovered a new pathway for evolution and genetic control.
Terry Clark, University of Sussex, 1985–91	Pioneered the study of macroscopic quantum objects.
Steve Davies, University of Oxford, 1985–91	Developed small artificial enzymes for chiral selection.
Nigel Franks et al., Universities of Bristol and Edinburgh, 1990–93	Quantified the rules determining distributed intelligence in animals.
Herbert Huppert and Steve Sparks, Universities of Cambridge and Bristol, 1983–92	Pioneered the new field of geological fluid mechanics.
Jeff Kimble, University of Texas at Austin and Caltech, 1983–92	Pioneered squeezed states of light.
Graham Parkhouse, University of Surrey, 1985–91	Developed a novel theory of engineering design relating performance to shapes and materials.
Alan Paton, Eunice Allan, and Anne Glover, University of Aberdeen, 1982–92	Discovered a new symbiosis between plants and bacteria.
John Pendry, Imperial College of Science and Technology, 1982–91	Pioneered the study of metamaterials.
Martyn Poliakoff, University of Nottingham, 1988–91	Transformed green chemistry: Supercritical fluids.
Ken Seddon, Queen's University of Belfast, 1988–91	Transformed green chemistry: Ionic liquids.
Colin Self, University of Newcastle, 1980–91, 1990–93	Discovered amplified enzyme-linked immunoassay; discovered that antibodies *in vivo* can be activated by light.
Gene Stanley and José Teixeira, Boston University and Laboratoire Léon Brillouin, 1990–93	Discovered a new liquid-liquid phase transition in water.
Harry Swinney et al., Universities of Texas and Bordeaux, 1985–91	Developed the first laboratory chemical reactors to yield *sustained* spatial patterns.

Table C: Venture Research Honors

Name	Honor
Mike Bennett, Plant Breeding Institute, 1986–92	OBE (1996)
Peter Edwards, Unversity of Oxford, 1987–92	FRS (1996)
Anne Glover, University of Aberdeen, 1982–92	CBE (2009), DBE (2015), FRS (2016)
Herbert Huppert, University of Cambridge, 1983–92	FRS (1987)
Jeff Kimble, University of Texas at Austin and Caltech, 1983–92	NAS (2002), Herbert Walter Award (2013)
Robin Milner, University of Edinburgh, 1981–87	FRS (1987), Turing Award (1991), Milner Lecture, Royal Society (2014)
J Strother Moore, University of Texas, 1983–86	NAE (2007)
John Pendry, Imperial College of Science and Technology, 1982–91	FRS (1984), Dirac Prize (1996), KB (2002), Kavli Prize (2014)
Gordon Plotkin, University of Edinburgh, 1981–87	FRS (1998)
Martyn Poliakoff, University of Nottingham, 1988–91	FRS (2002), CBE (2008), Royal Society Leverhulme Medal (2010), Russian Academy of Sciences (2012), KBE (2014), FREng (2017)
Ken Seddon, Queen's University of Belfast, 1988–91	OBE (2014)
Steve Sparks, University of Bristol, 1983–92	FRS (1988), CBE (2010), Vetlesen Prize (2015), KB (2018)
Gene Stanley, University of Boston, 1990–93	NAS (2004), Boltzmann Medal (2004)
Harry Swinney, University of Texas, 1985–91	NAS (1992), Boltzmann Medal (2013)

Preface

The genesis of this book began in December 2005 when a US National Science Board Task Force was meeting in Santa Fe, New Mexico, to discuss transformative research. Nina Fedoroff, the Task Force chairperson, suggested over lunch that I might like to consider writing an essay on how an organization might go about setting up a transformative research initiative. I have somewhat extended that remit, but I hope that the book also goes some way to meeting Nina's original specification. I am also immensely grateful to her for the invitation to join the Task Force and to fully participate in its extensive deliberations. Other Task Force members, Michael Crosby, Douglas Randall, and Jerry Pollack, were also especially helpful.

I am most grateful to Claudio Vita-Finzi, fellow founder-member of the "never say die" club, and an enthusiastic supporter over the years. John Allen, Peter Cotgreave, Irene Engle, Ross Gayler, Nigel Keen, Iain Steel, Ken Seddon, and Isa Zalaman have been unfailing sources of advice and much appreciated encouragement. Nina Fedoroff, Claudio Vita-Finzi, and Isa Zalaman also made many helpful comments on the early drafts for which I am grateful.

I would also like to express my gratitude to David Price, Duncan Wingham, and my many other friends in the Earth Sciences Department at University College, London, where I am a visiting professor. My weekly visits to the department's research seminars are invariably stimulating, and for a few hours each week they make it possible for me to pretend that I am a normal academic.

As ever, I am grateful to my wife, Shirley, and to David Braben, Peter and Lisa Braben, and Jenny and David Lightfoot, who have provided invaluable feedback in addition to their usual love and affection.

Donald W. Braben
July 2007

Scientific Freedom:
The Elixir of Civilization

—

Donald W. Braben

— Introduction

New products, new industries, and more jobs require continuous additions to knowledge of the laws of nature, and the application of that knowledge to practical purposes. . . . This essential, new knowledge can be obtained only through basic scientific research.

Science can be effective in the national welfare only as a member of a team, whether the conditions be peace or war. But without scientific progress no amount of achievement in other directions can insure our health, prosperity, and security as a nation in the modern world.

—Vannevar Bush
Science—The Endless Frontier, Report to the US President, 1945, p. 5

Thinking has always been humanity's greatest strength. That abstract ability separates us from the rest of the animal kingdom and has brought us to our present dominance. Like skipping in children, it is innate; it does not have to be taught. Civilizations have prospered or failed as our thirst for knowledge has been tolerated or suppressed. Remarkably, however, until the European Renaissance humanity's progress had been glacial, and centuries might pass before a detached observer might have noticed significant changes to global population or ways of life. The Renaissance then stirred the human spirit, and later created the conditions that launched the Industrial Revolution, first in Britain, and then rapidly elsewhere. Slowly at first we began to harvest our understanding of Nature and to use it for the greater good—material and intellectual. The result was that by the end of the twentieth century the average productivity of every man and woman had increased more than a hundredfold in real terms since the Renaissance. Global population also increased rapidly, but material wealth in the industrialized nations more than kept pace.

This prodigious progress came from our growing ability to harvest the fruits of humanity's intellectual prowess—scientific endeavor, as it is usually called. Material wealth continued to accelerate through most of the last century despite financial crashes and global wars. But then gradually, around about 1970, signs of major change began to emerge. Science's very success had unsurprisingly led to a steady expansion in scientists' numbers. That could not continue indefinitely, of course, and the inevitable crunch came when there were more than could adequately be funded. This was not only a numbers problem—the unit costs of research were also increasing. The funding agencies should have seen this coming, but they did not. Indeed, as I shall explain, many today do not accept this version of events, and are thereby contributing to one of the greatest tragedies of modern times. This perhaps surprising statement arises because the agencies' virtually universal response to the crisis was to restrict the types of research they would fund. Thus, to use a truly horrible word, they would *prioritize*, and focus funding on the most attractive objectives—that is, objectives the agencies *perceived* to be the most attractive. Thus, for the first time since the Renaissance, the limits of thinking began to be systematically curtailed.

The new policies would seem to have been phenomenally successful.[1] Modern life is enriched by vast and expanding ranges of astounding technologies. Communications, entertainment, food production, leisure, travel, and many other aspects of modern life have been transformed. Closer inspection would reveal, however, that most of this bounty stems from generic scientific discoveries made decades ago, a source that seems to have dried up in recent years. Consequently, our intellectual account is becoming overdrawn at a time when the demands on it are increasing. There is no shortage of initiatives aiming to deal with such problems as global warming, population growth, and terrorism, but one vital factor is usually overlooked.

Imagine for a moment that you are invited to list humanity's basic material needs. The list would probably contain the obvious things, such as food, water, heat, light, health, and security, but it might fail to mention the air we breathe simply because that vital ingredient can usually be taken for granted. Scientific freedom could indeed be placed in that latter category before 1970 or so because research policies then were usually based on laissez-faire. However, we have moved on, as they say. Nowadays, interference is the norm. However, as few funding agencies seem to have noticed, if current policies had applied at the beginning of the twentieth century, say, the world would now be a much harsher place. Today, far from being an inconvenience confined to feather-bedded academics, their consequences are approaching "the operation was a success but the patient died" category, and will affect the very foundations of our civilization.

---

1   An analysis was presented in my *Pioneering Research: A Risk Worth Taking* (2004).

Such abstract qualities as freedom are difficult or impossible to define. Freedom's loss may be easier to recognize, but it does not necessarily lead to chains. Increasingly nowadays, freedom is a *managed* commodity, but the consequences are subtle and varied. Indeed, at least for the time being, it is possible for almost everyone to live happily and productively within the current bounds. That is also generally the case in the sciences except for one essential factor. Those exceptionally rare scientists whose revolutionary work can open new horizons can do so only if they have total freedom. The routes to new types of knowledge can be deceptively disguised, and may appear to ordinary mortals as unimportant byways leading nowhere. There must be no filters whatsoever on what these scientists do, therefore, however well intended. Furthermore, their work is vital to future prosperity. In an increasingly complex and populous world, any attempt to limit it will lead us down the path to stagnation and pain.

Scientific progress comes in a vast number of ways, ranging from the apparently spontaneous comprehension of a new facet of Nature's behavior as typified by Albert Einstein's research on relativity,[2] say, to the prolonged and often agonizing study of a perplexing phenomenon as typified by Max Planck's work that led to the discovery of energy quantization. But if discoveries are to become part of the scientific lexicon, they must be endorsed by the scientific community, and that can often be problematic. However, leaving social problems aside for the moment, research for most scientists is indeed 99% perspiration with maybe 1% flashes of inspiration that hopefully culminate in the complex pieces coming together to form a coherent picture, at least in the investigator's mind.

Max Planck was one of the most influential scientists of the twentieth century, and therefore an appropriate role model for the story recounted here. In 1933, he wrote a typically succinct comment on the problems facing researchers who perceive serious flaws in accepted wisdom and know what to do about them:

No doctrinal system in physical science, or indeed perhaps in any science, will alter its content of its own accord. Here we always need the pressure of outer circumstances. Indeed the more intelligible and comprehensive a theoretical system is the more obstinately it will resist all attempts at reconstruction or expansion. And this is because

---

2   I have long used the word "Nature" as if to describe a being. My use has no religious or mystical significance, but is shorthand for the universe and every aspect of everything in it. We have made some progress, but our understanding of that infinitely faceted system is still in its infancy. We are not casual observers, of course, as Nature embraces the entirety of our very existence. Nature is, therefore, my affectionate and respectful anthropomorphism for a system that is the constant preoccupation of every scientist.

in a synthesis of thought where there is an all-round logical coherence any alteration in one part of the structure is bound to upset other parts also. For instance, the main difficulty about the acceptance of the relativity theory was not merely a question of its objective merits but rather the question of how far it would upset the Newtonian structure of theoretical dynamics. The fact is that no alteration in a well-built synthesis of thought can be effected unless strong pressure is brought to bear upon it from outside. This strong pressure must come from a well-constructed body of theory which has been firmly consolidated by the test of experimental research. It is only thus that we can bring about the surrender of theoretical dogmas hitherto universally accepted as correct. (Planck 1933, 44–45)

These words are as valid today as when Planck wrote them a lifetime ago, and reveal something of the dilemma he faced when he began his remarkable career some 50 years earlier. But the issues he identifies are still relevant. Indeed, they seem invariant and timeless.

One would hope, therefore, that research-funding organizations would have Planck's thinking in mind whenever they were contemplating new policies. Until about 1970, that effectively would seem to have been the case.[3] However, one must always be careful with generalizations. Not surprisingly, there is a voluminous literature, but it is concerned mostly with the qualities of freedom enjoyed by the profession as a whole. It deals with rights to freedom and academics' responsibilities to their various sponsors—society, government, industry, charities, and philanthropists—but the extent to which *individuals* can acquire or lose freedom is rarely discussed. Change in the academic world is often slow, so that my cited date is the peak of a broad distribution. Before 1970 or so, tenured academics with an individual turn of mind could usually dig out modest sources of funding to tackle any problem that interested them without first having to commit themselves in writing. Afterward, unconditional sources of funds would become increasingly difficult to find. Today, they are virtually nonexistent.

The way forward for ambitious young researchers was once clear, therefore. All they had to do was to acquire the necessary qualifications, and then to find a tenured appointment. To say the least, that was not easy, but not substantially more difficult than it would be today. However, having served their apprenticeship, they were free. They may have had to overcome the inevitable peer pressure if their plans were controversial,

---

3   Leon M. Lederman, when he was president-elect of the American Association for the Advancement of Science, said that funding for academic research was adequate in 1968, which year he described as the peak of a "golden age" of American science (Lederman 1991).

but *their peers did not have power of veto*—see **Poster 1**. Written applications were necessary if expensive equipment or large teams were required, but tenured researchers with modest needs would meet few obstacles. One's dedication and talent would usually be sufficient to silence the critics if the problem chosen were controversial, or if progress appeared to be lacking.

Poster 1
—

Charles Townes and the Laser

Charles Townes was awarded his PhD in physics at the California Institute of Technology in 1939, and went on to join the Bell Laboratories, then located in Greenwich Village on Manhattan Island. Soon after, the Bell management directed him to help develop radar-guided bomb-aiming systems as part of the US war effort. This intense work on radar and microwaves, as he describes it, led to his career's work on molecular spectroscopy. Similar war work had been done at the nearby Columbia University in New York, so when at the end of the war Bell suggested, as they say, that he should focus his work on subjects of interest to the company, he decided in 1948 that he would pursue his own interests, and accepted an appointment as associate professor of physics at Columbia.

For some time, he had been trying to make intense beams of submillimeter radiation, rather than the centimeter or more wavelengths that he had been working with. Eventually, he conceived a possible method to generate photon "avalanches" using excited ammonium molecules. But he could not get it to work. As he relates in his book:

> [After] we had been at it for two years, [Isidor Isaac] Rabi and [Polykarp] Kusch, the former and current chairman of the department—both of them Nobel laureates for work with atomic and molecular beams, and both with a lot of weight behind their opinions—came into my office and sat down. They were worried. Their research depended on support from the same source as did mine.[4] "Look," they said, "you should stop the work you are doing. It isn't going to work. You know it's not going to work. We know it's not going to work. You're wasting money. Just stop!" (Townes 1999, 65)

---

4   Their support came from a Joint Services contract managed by the Army Signal Corps.

But Townes had come to Columbia on tenure, so he knew he couldn't be fired for incompetence or ordered around. Nevertheless, the awesome weight of Rabi's reputation in particular—a one-time senior member of the Massachusetts Institute of Technology's legendary Radiation Laboratory, set up by Vannevar Bush to develop wartime radar—must have been daunting. Such top brass cannot be defied lightly, and showing extraordinary courage, this junior faculty member stood his ground, and respectfully told his exalted colleagues that he would continue. Two months later (in April 1954), his experiment worked, and the maser (microwave amplification by stimulated emission of radiation) was born. Three years after that Arthur Schawlow, Townes's postdoc at Columbia, had moved to the Bell Laboratories, and their collaboration led to the optical version of the maser—the laser. Townes was awarded the Nobel Prize in Physics in 1964 for these discoveries (shared with Aleksander Prokhorov and Nikolai Basov [USSR], who developed the maser and laser independently). Schawlow was awarded the Nobel Prize for Physics in 1981 for his work on laser spectroscopy.

As Planck says, researchers who claim that an accepted dogma is seriously flawed have a duty to persuade the scientific community that their alternative is better. Planck was indeed his own severest critic in this respect, as serious researchers often are. As he would have been the first to point out, the status quo should not be changed lightly. He was, in his own words, a reluctant revolutionary. In Planck's time, academic research was essentially unmanaged, whether by objectives or otherwise. There was no central direction or coordination. The issues involved were largely scientific. That is no longer the case. Now that the new policies are firmly established, for the first time in science's long history researchers must submit their proposals in writing to an appropriate agency. Spontaneity has been lost. The funding agencies subject these proposals to an arcane set of tests (peer review) designed to flag what they perceive as the best, expecting thereby that the rest will probably be lost.

These well-intentioned changes have created lumbering bureaucracies to ensure compliance. They have also inhibited exploration outside the mainstreams and challenges to convention. This is most unfortunate because the great discoveries that transformed the twentieth century came out of the blue. *There was no demand for them.* One might suggest that the Manhattan Project and the Human Genome Project are among examples to the contrary, but the unexpected discoveries on which they were based had come much earlier. Their successes are monuments to organized creativity and depended on exquisite fine-tuning of existing knowledge and orchestration of resources on unprecedented scales rather than on new science per se. However, scientists today with radical turns of mind—the successors to Planck et al. (see **Poster 2**)—are unlikely to get funded because their ideas are unlikely to impress a committee *before* they have been confirmed. Consequently, there has been a dearth of major scientific discoveries in recent decades. We are living off the seed corn.

Poster 2
—

The Twentieth-Century Planck Club

The twentieth century was strongly influenced by the work of a relatively small number of scientists. A short list might include Planck, Einstein, Rutherford, Dirac, Pauli, Schrödinger, Heisenberg, Fleming, Avery, Fermi, Perutz, Crick and Watson, Bardeen, Brattain, Shockley, Gabor, Townes, McClintock, Black, and Brenner (see Table 1). However, I give this list only to indicate something of the richness of twentieth-century science. I wrote it in a few minutes, and it obviously has many important omissions. Other scientists would doubtless have their own. If the criteria for inclusion were based on success in creating radically new sciences, or in stimulating new and generic technologies, a fuller list could easily run to a couple of hundred.

I will refer to the extended list as the Planck Club or alternatively as Planck et al. for the remainder of the book.

However, issues much wider than research are at stake. In everyday life we rely, usually unquestioningly, on humanity's basic needs being available at affordable prices so that we can get on with our lives at our own pace. For those of us living in the industrialized countries—say, the 20 or so richest states of the Organization for Economic Cooperation and Development (OECD)—we can usually take them for granted. But global population is steadily increasing. In addition, many millions in China and India, for example, deemed hitherto not to have such needs are now beginning to assert their rights to them. Demand for some vital resources—oil, copper,[5] and water—is increasing rapidly. Many doomsayers take the

---

5   *The Economist*'s metal index published on March 22, 2007—which includes aluminum, copper, lead, nickel, tin, and zinc—was at its highest level for 16 years.

Malthusian line. Earth's resources are strictly limited, they say, and we are all going to hell in a handcart.

They are wrong, because they reckon without humanity's apparently boundless command of the intellectual dimension. Thanks to that precious gift, and despite the havoc of world wars, financial crashes, and a threefold rise in population, per capita economic growth soared in the twentieth century,[6] reaching a peak, coincidentally perhaps, around about 1970. It then began a steady decline. Unfortunately, however, that decline is now being masked by unprecedented rises in residential and other property values, dubbed by *The Economist* as perhaps the biggest bubble in history (see Chapter 3). Bubbles usually collapse more quickly than they inflate, and who knows what the effects will be. Humanity is indeed blessed with the ingenuity necessary for survival, but much of this priceless asset will be wasted if we smother it with consensus. See **Poster 3**.

This book seeks to extend the debate to anyone who takes a serious interest in global affairs—industrialists, academics, legislators, consumers. Sadly, however, even though the fruits of research are essential for modern life, the media seem to confine their interest to specific *discoveries*; research as an *enterprise* is generally ignored, and *research policies* have even less appeal apparently, if that is possible. This indifference would seem to have a simple explanation. Since the Renaissance, the policy on academic research has generally been *not* to have a policy. Patronage and sponsorship have always been important, but their agendas were diverse and they often backed creative talent for its own sake. Society's reward was a remarkably steady trickle of astonishing miracles. But times have changed; funding agencies' mission statements (or other expressions of purpose) are now de rigueur, and often cast in stone. This situation has long been the case for researchers in the industrial sector, but that is to be expected. Industrialists must know where they are going. Now, academics, too, are almost invariably subject to externally imposed constraints. Unfortunately, these dramatic changes seem to have passed the media by, so would-be reformers are starved of the oxygen of publicity.

---

6   During the twentieth century, the average real-terms productivity of every person on the planet increased 4.3-fold despite a threefold increase in population.

Poster 3
—

The Wizard's Warning[7]

Once upon a time, I dreamed about a meeting about a hundred years ago at which a wizard addressed a large gathering of industrial and scientific leaders. He announced that he would use his 20/20 foresight to describe the powerful discoveries that would enrich the coming twentieth century. "However, your language does not yet contain the words I need to describe the future," he said, so he put a spell on the audience that conjured visions of energy quantization, relativity theory, atomic and nuclear structure, quantum mechanics, and molecular biology to give some impressions of the sciences that might shortly come. With mounting excitement, he outlined some of the magical technologies that might stem from them: magnetic resonance imaging and a wealth of other medical diagnostics, lasers, nuclear power, computers, telecommunications, and genetic manipulation. The audience was quite literally spellbound. They had expected him to talk about developments in electrical and steam power, hydraulics, and oil and coal technologies. What they heard was totally unexpected. When the wizard had finished, someone asked what we had to do to make these fantastical things happen. His reply was equally stunning: "Nothing," he said. His voice began to quiver with emotion: "Humanity has been given the priceless gift of creativity, but it's vital that you understand how it works. Creativity is the essence of the human spirit, and flowers best when it's unconstrained. If you try to control it for your own ends you must learn that you can get only what you ask for. The unexpected will not arise. You are not wizards." These last words came in an intense growl. Then he disappeared, and I woke up.

I dreamed about the wizard again recently. He told me that he had given similarly prescient lectures every hundred years since 1600, when Francis Bacon and a few others began to appreciate the value of organized scientific research. Before that, the scientific world had been more or less stagnant for a long time, so there was nothing new to herald. His lecture on the twenty-first century was clearly long overdue, and I asked him when next he would be speaking to us. His answer was depressing:

> I will give no more lectures until humanity regains its sense of wonder. As you know, my foresight is perfect, but I'm not allowed to reveal

---

7   Based on an editorial by the author published in *The Scientist* on September 27, 2004.

anything about the major discoveries that await you. The sole purpose of my centennial addresses has been to inspire, to conjure for the best of you subconscious images of what your creative talents are currently capable of. The rest is up to you. Unfortunately, your leaders have now decided that wonder is inefficient because it cannot be controlled, quantified, or targeted. You must consolidate what you think you know, of course, but nowadays that is all you are doing. Humanity's powers of foresight have always been puny, so you'll get nowhere unless you listen to what I am trying to tell you and back those rare individuals whose vision transcends need.

Never say die, as my good friend says, so I will outline what we should do about our currently bizarre predicament and who should do it. For once, finance is not the major issue. The initial cause of the funding crisis was the mushrooming numbers of researchers and the high levels of inflation applying to scientific equipment, an inflation being pushed ever higher by the pressures arising from enforced competition. Today, however, no realistic increase in public funding could provide for every demand. Instead, funding agencies should inject fresh air into their somewhat stagnant thinking and break away from their slavish adherence to priorities and consensus. The financial cost of liberating a small proportion of the scientific enterprise—say, for transformative research initiatives—would be barely noticeable.

However, the rules by which governments and others oblige them to operate are too rigid. Accountability for every cent means that the return on every cent must be justified. The result is short-termism.[8] Special-interest groups in such fields as the environment or ethics usually seem to find effective ways of making their views known. Unfortunately, the collateral damage inflicted on the governance of science has generally gone unnoticed because the *recipients* of public and other funds are unlikely to complain, and for the rest, who takes notice of sour grapes? But poor harvests will affect us all, and we must find ways of raising the profile of these issues if we are to restore the vital role played by freedom in research.

Jared Diamond, in his *Collapse* (2005), tells the fascinating story of how civilizations choose to fail or survive. His study covers many centuries and explores the sudden collapse of a wide variety of small or localized

---

8   I discussed these problems with Akio Morita when he was Sony's chairman some years ago. He replied with a story about an ocean liner that had just hit the rocks. The captain and his senior officers were playing poker in his cabin. An engineer knocked on his door with news that the ship was sinking fast. "Go away," said the captain. "Can't you see we're playing for high stakes?"

societies that were "especially clear illustrations," as he puts it, of the problems that may befall larger societies today. Easter Island is perhaps the best known.

Polynesians arriving on this remotest of Pacific islands around 900 AD found it hospitable and covered in dense forest. However, drawing on a wide range of archaeological and forensic science, Diamond concludes that the forest finally disappeared by about 1600 to provide the most extreme example of forest destruction in the Pacific.

This deforestation came about largely because the islanders needed trees for the construction and transport of an escalating number of huge stone statues that played an essential role in the islanders' religious and social lives. Unfortunately, the island's ecology is totally dependent on trees, and demand for them eventually outstripped Nature's supply. Without trees, migrating birds could not nest and a vital food supply vanished. Without trees, the islanders could not build long-range canoes, and so deep-sea fish, dolphins, porpoises, and tuna became inaccessible. Loss of trees also led to increased soil erosion and poor crop yields. Diamond estimates that by 1774, when Captain Cook visited the island, its once prosperous population had declined by about 70% since its peak between 1400 and 1600, and relates that Cook described the islanders as "small, lean, timid, and miserable." What had happened? We do not know for sure, of course, but Diamond asks (on p. 426): "What did the Easter Islander who cut down the last palm tree say as he was doing it?"

Diamond goes on to explain the fate of other civilizations, and to discuss the relevance of his story to humanity's survival today. He concludes (on p. 522) with some reasons for hope:

> We don't need new technologies to solve our problems; while new technologies can make some contribution, for the most part we "just" need the political will to apply solutions already available. Of course, that's a big "just." But many societies did find the necessary political will in the past.

Diamond rigorously covers the material aspects of his story, but in my opinion he does not deal with its intellectual dimensions with the same meticulous attention to detail. For Easter Island, for example, it is likely that one or two dissenting voices would always have been railing about the madness of trying to cover the island with statues. But who would have listened? Statue building was apparently the principal mode of competition among the island's dozen or so factions. If one faction withdrew from the statue-building race, it would effectively be throwing in the towel. The leadership might have pointed out that the island still had a viable number of trees and a much larger number of saplings. We do not have to worry, therefore, their leaders might have said, the saplings will grow in due course and more than replace the trees we take. Without a forum and with no power base, the dissenters would have been marginalized and ignored.

In other circumstances, the leaders might have been right, but unfortunately the island is normally subject to very strong winds. An exceptionally strong gale could destroy the entire tree cover if it had already been seriously depleted, and the saplings would suddenly be left with no cover and soils reduced by erosion. Tree cover would most likely have come to a catastrophic end, therefore, rather than dwindling away imperceptibly as Diamond surmises. My conjecture is that their actions had caused the island's ecology to become precariously and uncontrollably balanced; that is, they had pushed it into what we might call the Damocles Zone (see Chapter 1). Once tree cover had fallen below a certain critical level, the islanders' fate would have been sealed. They could then have done little or nothing to prevent it. I cannot offer hard evidence, of course, but if competition is paramount, how could the islanders have mustered the political will to solve their looming problem before it became too late?

Is Diamond's sanguinity justified? The political will to "apply solutions" is an essential condition, of course. However, full confidence would be justifiable only if actions were based on a *coherent* understanding of all the political, social, and scientific inputs involved. Otherwise, actions would be no more effective than wishful thinking. In any case, the political will alone may not be sufficient, especially if we defer essential action on crucial problems for so long that the conditions leading to collapse *become impossible to avoid*. With a coherent understanding, every conceivable contributing factor will have been identified and hopefully, therefore, we can develop the antennae to give early warning of possible problems and the flexibility required for those that take us by surprise. **Poster 4** gives an illustration. The signals may not be to everyone's liking, of course. Indeed, that will rarely be the case. But unlike my imagined dissenters on Easter Island, we fortunately do have a locus for those capable of seeing the writing on the wall and diagnosing solutions. It is called the *university*. For many centuries, that institution has often defied the establishment and has always stood ready to offer guidance and advice, but the recent changes undermine its independence and potential.

It is possible that my message may be seen as elitist and of interest only to those very few scientists who might be putative members of a twenty-first-century Planck Club. That interpretation would be wrong. One of the themes here is that almost every serious researcher is at some time in a career capable of taking those fateful steps that might lead to a great discovery or the creation of penetrating new insight. They might then need to draw on vast reserves of courage and determination, and perhaps also a little luck if they are to make progress. At any one time, of course, the proportion of researchers ready to seize that possibly once-in-a-lifetime opportunity will be very small, so if they are prevented from doing so the democratic pressure they can exert is insignificant. This is becoming a huge problem. It will only be solved if the beneficiaries—that is, the general public—take an interest and do something about it. It is their, that is *your* future that is at stake.

Poster 4
—

Global Warming: A Coherent Approach

The atmosphere is host to a wide range of complex processes that a large number of climate modelers worldwide are struggling to understand. However, it is not an isolated entity. One boundary is in intimate contact with a similarly complex but much larger body—Earth—that is about a million times more massive. The other separates us—humanity—from the rest of the universe. Thus arises one of the twenty-first century's most baffling questions: Are our activities affecting the planet?

The atmosphere's first terrestrial contact is largely with the oceans—300 times more massive and covering about two-thirds of Earth's surface. Transport of heat, water, gases, and particulates across the ocean–atmosphere interface plays a key role in the global climate system. The oceans transport huge quantities of heat—from the equator to the poles, for example—and play crucial roles in the global carbon cycle. Biological effects are also important. Marine microbes are responsible for about half of Earth's production of carbon and nutrients such as phosphorus and nitrogen, and their cycling can affect atmospheric concentrations of CO_2. As a further illustration of these complexities, in 2006 it was unexpectedly discovered that the migration of marine animals may make important contributions to global warming (Kunze et al. 2006). Biologically generated turbulence can significantly affect the transport of heat and nutrients, and air–sea gas exchanges. We understand few of these complex processes.

Thermal processes within Earth itself are highly complex. Earth's magnetic field originates in convection currents within the outer (molten iron) core. The radius of the inner, solid-iron core seems to be increasing. Thus, there are liquid–solid phase changes that necessarily involve the exchange of large amounts of latent heat of which we understand very little. The temperature of the core, for example, is uncertain to about 1000°C. Earth's magnetic field has declined about 20% in the last hundred years, and it is possible that we are seeing the prelude to one of the periodic magnetic-pole flips that occur on average about every million years. The atmosphere is therefore being subject to unknown but probably substantial fluctuations in its heat exchanges with Earth. Their timescale may be long, perhaps longer than the effects we see in the atmosphere, but we simply do not know.

For the atmosphere's upper boundary, the Sun also changes its long-term output (years to decades) in ways that we do not understand. Indeed, it is possible that these solar fluctuations can account for the observed changes in Earth's temperature over the past 100 years (Friis-Christensen and Lassen 1991). Furthermore, our planet is bathed with solar and galactic cosmic rays that vary in intensity unpredictably, and to compound these problems, the strength of its protective magnetic field is falling. Cosmic rays can affect the structure of the upper atmosphere, and lead to changes (cloud formation rates, etc.) that in turn can affect climate. No doubt there are other problems of which we are as yet unaware.

The processes that determine the Earth atmosphere's behavior are nonlinear; see Chapter 1. Feedback mechanisms are therefore essential to stability. As Earth warms, for example, evaporation of water vapor—another greenhouse gas—should increase, thereby increasing global warming, but that also should increase the rate of cloud formation. Clouds increase Earth's reflectivity, and hence reduce the Sun's heat arriving at the surface. But the delicate balances involved in these processes are unknown, as is their relative importance compared with ice-cap melting, say, which should decrease reflectivity. There are in fact a multitude of such feedback mechanisms, and most, if not all, of them are poorly understood.

However, climate modelers are making heroic attempts to tame these complexities and make predictions, but computer models can only be as good as the data they use. In addition, long-term predictions of nonlinear systems can be fraught with danger. Although there are many models of world economic performance, for example, and human interactions are reasonably well understood, how many modelers would dare to try to predict inflation levels, say, in 2050?

There is an urgent need, therefore, for coherent data rather than, as one hears so much today, calls for more powerful computers. The ozone hole, for example, was discovered by the British Antarctic Survey in 1985 when Joe Farman and his colleagues suspected that computers were automatically rejecting vital data that on current understanding were deemed impossible but that nevertheless were evidence for the hole (Farman et al. 1985). In these circumstances, the only viable approach is to encourage unfettered scientific exploration. What we have, unfortunately, are armies of explorers obliged to treat components of this hypercomplex system as though they can be studied in isolation—as if each component were independent of all the others. Coherent approaches are unlikely to be funded because they would cross too many conventional disciplinary boundaries.

This does not mean that I advocate inaction. For example, in the absence of understanding, it would be irresponsible to continue injecting large and rising amounts of greenhouse gases such as CO_2 into the atmosphere.[9] They should be reduced, of course, or at least not increased until we are reasonably sure that we know what we are doing. What we should not do, however, is to claim that if humanity takes such actions, the global warming problem will be solved. It might not. The main causes might lie elsewhere, or there might not even be a problem.

The book is structured as follows. **Chapter 1** introduces the formidable Damocles Zone, and outlines how we can avoid it. The Sword of Damocles is, of course, a fable, but the consequences of moving into that eponymous Zone can be all too real. That is because, despite the signs to the contrary, the world is very strange. Virtually every process on which life is based and indeed every dynamic process in the air, earth, oceans, or the universe itself is ultimately finely balanced. Collapse and instability are never far away. However, for most of us, and for most of the time, that is not what we actually observe. The world is generally a pleasant and predictable place only because of the feedback mechanisms that usually tame the instabilities. My conjecture is that at the highest social levels creativity provides the vital feedback that keeps societies and civilizations healthy. Crushing it, as was done during the Dark Ages, for example, pushes us into that fateful Damocles Zone and makes us highly vulnerable. Today, our current obsession with so-called efficiency and accountability is hardly consistent with promoting creativity, but in sharp contrast to the Dark Ages we have expanding populations with increasing expectations.

We *must* take more care or suffer the consequences.

---

9   An excellent summary of the problems associated with controlling or trading carbon emissions is given in an editorial written by Schlesinger (2006).

Chapter 2 describes transformative research, the term now used in the United States to describe the research that has a reasonable chance of radically changing our understanding of important concepts or creating new fields of science. I avoid a more precise definition, however, because the main objective of any transformative research initiative should be to rectify the serious flaws in current funding arrangements and thereby stimulate the creation of a twenty-first-century Planck Club. The major snag is that every discovery made by their twentieth-century predecessors was unexpected. That not only creates difficult acts to follow, but sponsors of these initiatives must recognize that major breakthroughs in scientific understanding are impossible to commission. The most we can do is to restore the freedom that leads to them.

Chapter 3 discusses why we should be concerned about our current predicament. Global per capita economic growth rates are falling, a fact that seems related to the policy changes made since about 1970. So-called breakthrough discoveries made in recent times are contrasted with some of the pre-1970 vintage that so transformed our lives. Before that fateful date, scientific discovery was intensely personal. Nowadays, however, research tends to follow the fashions of consensus. Large numbers of developments are attributed impersonally to anonymous "researchers" rather than to highly individualistic Olympians such as Planck, and most of these are likely to be superseded by the next development before too long. I attribute the apparent fall in scientific productivity to the loss of freedom in research, but the loss is not irrevocable. Scientists willingly gave up their freedom during World War II to serve their countries, and their contributions in such areas as radar, communications, code breaking, and the atomic bomb were crucial. That directed management-by-objectives mode could easily have become the norm as it was so obviously successful, but thanks to the actions of a few determined people, the peace brought a range of brilliantly effective measures to restore academic freedom. However, their very success led to an unprecedented expansion of scientific enterprise, and consequently to demands for funds outstripping supply. We are therefore in dire need of a few determined individuals who can mount effective challenges to the new policies. It is never easy to accomplish change, but maybe they can take their inspiration from kindred spirits of the late 1940s.

Chapter 4 discusses how we might go about finding Planck's successors. The essential point is that such initiatives must pass the "Planck test" if they are to stand a chance of succeeding. Thus, search parties must genuinely believe that their procedures would probably have led to the support of Planck et al. *when they were starting out*. Unfortunately, the funding agencies seem preoccupied with so-called high-risk research. They do not seem to have realized that they have hijacked the management of risk—that should normally be the researcher's responsibility, at least for academics. For our part, we modeled our 1980s Venture Research initiative on the work of Planck Club members, and we could not believe that they thought their work was high-risk. Judging by the high rate of Venture Research successes, it seems we got it about right.

Chapter 5 describes how a transformative research initiative might work in practice. Unfortunately, one's procedures must also be transformative if the effects of the post-1970 changes are to be mitigated. The ultimate objective is to create a twenty-first-century Planck Club, of course, but likely members of such an august Club will probably be very careful when sharing their ideas. There must be mutual trust, and that must be won. Indeed, the staff of a transformative research initiative has a very special role to play. Every member of the twentieth-century Planck Club enjoyed freedom as a right, but in its absence, staff must try to be *Nature's ambassadors* and act as Nature herself might if she had to make the selections.

Chapter 6 discusses the university as an institution, the very home of the intellectual dimension. This most successful of our institutions has served society well over the centuries. In the twentieth century, it provided an encouraging environment for most members of the Planck Club (but not all; some were from industry) and many others who made vital contributions during World War II. This extraordinary reservoir of talent is now under serious threat. The objective of offering higher education to 40–50% of the student cohort may be desirable, but unless we take steps to preserve, for example, the freedom of research, the university as we have known it will disappear, having drowned in a sea of mundanity. Possible solutions are discussed, but they would seem to require a revolution—a proposed Fifth Revolution—the natural successor to the generally accepted four that have dominated our lives for the past 300 years or so. Transformative research initiatives are unlikely to succeed unless we make some progress toward bringing that revolution about.

Chapter 7 gives the first full account of the 26 programs being funded by the Venture Research initiative at its close in 1990, and of their eventual progress. No programs are excluded. As will be seen, Venture Researchers were phenomenally successful. At least 14 groups or individuals made transformative discoveries, but their proposals had all been rejected by the usual funding agencies. These high success levels would seem to indicate that Venture Research (transformative research, by another name) is not high-risk, and could be a viable alternative, at least for radical researchers, to the environment prior to the 1970s or so that generally allowed freedom for all. It is also remarkably cheap.

Humanity's future seems precariously balanced. We are facing a daunting catalog of problems of which perhaps the most intractable is that the majority of the world's population are now demanding the higher standards of living long enjoyed by the industrial nations. There is no reason in principle why that cannot happen. However, if we are all to have decent shares of cake, there must be a lot more cake; that is, global per capita economic growth must continue to increase, year after year. Unfortunately, the underlying trend in recent years seems to indicate a decrease. Growth depends on technical change, but management by objectives and other instruments of bureaucracy are strangling scientific research, and our universities are struggling to cope with their rising demands. Thus, the global warning seems clear—falling growth and rising expectations could drive some regions into a chronically unstable state leading to the Damocles Zone and collapse. We do not know when that might happen, but it makes no sense to assume that it will not. On the other hand, if we strive to ensure that the most original of our elite scientists and universities are completely free to tackle these and *any other* problems, experience tell us that we could synthesize the elixir that perpetuates civilization indefinitely.

1   The Damocles Zone

"My other piece of advice, Copperfield," said Mr. Micawber, "you know. Annual income twenty pounds, annual expenditure nineteen [pounds] nineteen and six, result happiness. Annual income twenty pounds, annual expenditure twenty pounds ought and six, result misery."

—Charles Dickens
David Copperfield, 1850, Ch. 12

Charles Dickens had no training in the sciences, but that is hardly a prerequisite for an understanding of what seems to be a fact of life: that a *sustained* shortfall in one's circumstances—and not only the financial, of course—can eventually lead to catastrophic consequences. It would seem that after lengthy immersion in such dire straits, those affected enter a state of existence in which even trivial setbacks such as an unexpected bill, a minor fall, a modest bout of food poisoning, a common cold, or freezing weather, which in normal circumstances would be taken in their stride, can lead to serious difficulties or even disaster simply because they have reached the end of their tether and can no longer cope. During that precarious phase it might seem to the disinterested observer that tiny causes can become amplified to produce disproportionately huge effects on its victims. Moreover, they seem to be powerless to do anything about it—that is, they will have entered the Damocles Zone (see **Poster 5**).

However, it is not only the poor unfortunates who might find themselves in such a predicament. It potentially awaits us all, not only as individuals, but as societies, nations, and civilizations, and perhaps extending to humanity itself. Passage into the Damocles Zone is not necessarily the result of transient mishaps such as storms, tsunamis, earthquakes, or even being caught in someone else's war. A healthy society should indeed be able to take these things in its stride. Their impact on individual victims can be disastrous, of course, but experience shows that societies generally recover remarkably quickly. Such disasters can arise out of the blue and may come about as accidentally as a car crash with few implications for humanity as a whole. The Damocles Zone has wider and deeper origins, however, and its effects would only rarely be reversible.

Poster 5
—

The Sword of Damocles and the Damocles Zone

The Roman politician and philosopher Marcus Tullius Cicero, writing in his *Tusculan Disputations*, Book 5, tells a story about King Dionysius II of Syracuse and the Sword of Damocles. He says (translated by C. D. Yonge, 1877):

> XXI. This tyrant, however, showed himself how happy he really was; for once, when Damocles, one of his flatterers, was dilating in conversation on his forces, his wealth, the greatness of his power, the plenty he enjoyed, the grandeur of his royal palaces, and maintaining that no one was ever happier, "Have you an inclination," said he, "Damocles, as this kind of life pleases you, to have a taste of it yourself, and to make a trial of the good fortune that attends me?" And when he said that he should like it extremely, Dionysius ordered him to be laid on a bed of gold with the most beautiful covering, embroidered and wrought with the most exquisite work, and he dressed out a great many sideboards with silver and embossed gold. He then ordered some youths, distinguished for their handsome persons, to wait at his table, and to observe his nod, in order to serve him with what he wanted. There were ointments and garlands; perfumes were burned; tables provided with the most exquisite meats. Damocles thought himself very happy. In the midst of this apparatus, Dionysius ordered a bright sword to be let down from the ceiling, suspended by a single horse-hair, so as to hang over the head of that happy man. After which he neither cast his eye on those handsome waiters, nor on the well-wrought plate; nor touched any of the provisions: presently the garlands fell to pieces. At last he entreated the tyrant to give him leave to go, for that now he had no desire to be happy. Does not Dionysius, then, seem to have declared there can be no happiness for one who is under constant apprehensions?

Following Cicero's thought-provoking story, I will use the term *Damocles Zone* to describe that state within which its occupants are suddenly exposed to grave and imminent danger. That might be the case even though the obvious signs have been benign until that moment. On moving into the Damocles Zone, not only does one's fate become precariously balanced but also survival is probably impossible without external intervention. Although a disastrous outcome is virtually certain, no estimate can be made as to when it might actually happen.

The existence of Damocles Zones of one kind or another is an intrinsic property of the universe we live in.

If the universe were dominated and controlled by linear processes we would always be safe from risk, accidental or otherwise. By *linear processes* I mean any process in which the consequences of an action are predictable and directly proportional to its causes. I might push you in a friendly way, say, with a force that might normally move you a few inches, and in a linear world you would indeed move only a few inches. On the other hand, if you were standing at the edge of a cliff, you might go over the edge and suffer consequences that would probably be nonlinear and unpleasant. But if linear processes ruled the world, there would be no cliffs. Cliffs are always the result of some discontinuous and perhaps tumultuous event. In a linear world, all surfaces would either be flat or very gently undulating. Skiing would be out because without tumultuous events there could be no mountains. But that would be the least of our worries for another important reason—in a linear world not only would we be safe from all risks but there would also be no people and probably no world, either.

Our understanding of the universe is rather limited, of course. However, one feature that seems well established is that the universe is shaped and determined entirely by *nonlinearities*; that is, the nonlinear relationships between every entity—space, time, energy, matter, and so on—of which the universe is composed. This prosaic and tongue-twisting word is therefore the source of everything that is interesting and exciting, that makes life worth living, and indeed that makes it possible to live. Eventually, nonlinearities always lead to instability. There seems little doubt that the universe's very creation, regardless of whether one accepts the Big Bang theory, comes from spontaneous instabilities from which matter emerges, and by which, on the scale of the universe, energy is conserved. Other types of instability seed the creation of individual stars, planets, and galaxies. The very chemical elements on which life on Earth is based (carbon, oxygen, nitrogen, etc.) come from yet another—the spontaneous explosion of stars (supernovas) that flood the universe with them. We have yet to discover precisely how life on Earth emerged, but we do know that proteins play an essential role in the later stages of that process, and that proteins are essentially unstable.[1] If they were not, we could not exist as there could be no living cells, no renewal through birth and death, not only of cells but of whole organisms, and no mutations and no evolution. Even when life on Earth became established, its later extensive development into higher forms became possible and sustainable only because of physical instabilities in the core of the planet. They led to the steadily expanding crystallization of the solid-iron core at the heart of Earth's hot interior, and in turn to the subsequent turbulent flows in the conducting outer core of molten iron that generate the magnetic field needed to protect higher life-forms such as ourselves from much of the harsh and damaging radiation from the Sun and beyond. Thus, without nonlinearities we would not only all be dead—we would never have lived.

That brief tribute to instability, however, omits a very important caveat. Although nonlinear processes would seem eventually to dominate every system ever observed, the universe at any one time is nevertheless a curious mixture of the linear and the nonlinear. Instability must be followed by a period of stability otherwise nothing new would survive. Unfortunately, these considerations take us into one of the most profound of all the sciences that at least on the astronomical scale is poorly understood—thermodynamics. Thus, the universe also seems to be characterized by the property that an ordered system evolves in such ways that it always becomes less ordered—that is, entropy must always increase as the arrow of time goes only one way. (As far as we know, the direction of that arrow never reverses.) Eventually, therefore, the universe will become an amorphous soup in which no structures can exist and nothing can ever happen—a bleak prospect, indeed.[2] But returning to present reality, it goes without saying that we do exist. Our immediate astronomical neighborhood also seems generally stable, and most of us can enjoy mostly quiet lives. Entropy's eventual absolute hegemony, therefore, can be understood only if it applies to the universe as a whole.

On this viewpoint, therefore, some systems might behave linearly for long periods provided that a sufficient number of other systems behave nonlinearly (thereby becoming less ordered) over the same time so that the entropy of the universe indeed always increases. That anthropic

---

1 Proteins are essential components of all living cells of all organisms. They may differ in sequence, shape, and function, but they must all be able to fold into specific three-dimensional structures. These structures are not rigid. They each have a restless and dynamic existence, which involves unfolding and refolding, complex association and dissociation.

2 The universe is expected to become more and more disordered until it finally ends up as a featureless goo, but according to present understanding, that depressing fate is scheduled after the passage of such a ridiculously long time—perhaps 10^{100} years—that it would seem only to indicate the extent of our ignorance.

thinking would seem to rationalize our existence nicely,[3] but while all this may be true, the mysterious forces harmonizing entropy's inexorable rise over astronomical timescales need not concern us here. On the everyday timescale, Nature *depends* on instability. Paradoxically, nothing could exist without it. In nucleic acids, the need for instability is well established—evolution depends on mutation and rearrangement of DNA. Without the ability to break down or to promote changes in the constituent elements of living systems, there would be no development. However, the complex processes that make up living systems such as humans, say, are all subject to rigorous control, for which, of course, there must be appropriate feedback mechanisms. A human, for example, is made of some 10^{14} cells in some 200 different types of tissue, and their behavior is linear—that is, predictable—for most, if not all, of their lives.

The quest to understand the requisite control mechanisms in biology is now a major scientific discipline. As Mathew Freeman explains in a recent review:

> The intercellular communication that regulates cell fate during animal development must be precisely controlled to avoid dangerous errors. How is this achieved? Recent work has highlighted the importance of positive and negative feedback loops in the dynamic regulation of developmental signalling. These feedback interactions can impart precision, robustness and versatility to intercellular signals. Feedback failure can cause disease. Negative feedback occurs when, for example, a signal induces the expression of its own inhibitor; it serves to dampen and/or limit signalling. Positive feedback occurs when a signal induces more of itself, or of another molecule that amplifies the initial signal, and this serves to stabilize, amplify or prolong signalling. (Freeman 2000)

Researchers generally seem to regard instability as a nuisance that they must deal with. However, instability is essential to Nature's purposes, and so researchers should also be able to use it to their advantage. Indeed, Colin Self's Venture Research work (see Chapter 7, VR 23) was dedicated to understanding its role in biology and particularly in the immune

---

3   As Richard Feynman and colleagues put it:
     For some reason, the universe at one time had very low entropy for its energy content, and
     since then the entropy has increased. So that is the way to the future. That is the origin of all
     irreversibility, that is what makes the processes of growth and decay, that makes us remember the past and not the future, remember the things which are closer to that moment in the
     history of the universe when the order was higher than now, and why we are not able to remember things where the disorder is higher than now, which we call the future. (Feynman et
     al. 1963, 45–46).

system. Nonlinearity is at the heart of all biochemical processes, which Nature ensures are regulated by the appropriate feedback controls. The degree of instability can therefore be understood as being related to the need for feedback. If a process were to persist for too long (i.e., if its instability were too low), the process would go out of control unless the requisite negative feedback were applied at the right time. Alternatively, a weak and transient response to a stimulus (i.e., if its instability were too high) would need to be amplified by positive feedback if the response were to be effective. When we accidentally cut ourselves, for example, we need not normally fear for our lives. This is because antigens invading from the outside world almost instantaneously trigger the activation of a blood-clotting agent that seals the wound so that eventually it can heal. The next step is equally crucial. The agent must then be switched off! If the agent had too little instability (i.e., if it were too stable), it would exert its effect for too long, resulting in our entire blood supply rapidly coagulating into a solid mass following even a minor accident. Conversely, if the agent were too unstable, its effect would be transient and ineffective, and blood loss would continue unabated. Nature gets the balance just right, of course, or we would not be here.

Indeed, life would appear possible only because of a truly vast number of exquisitely balanced controls, not only at the molecular and cellular levels but also at the macroscopic levels in all living organisms. Thus, for example, these controls maintain human body temperature at approximately 37°C independently of whether we are resting, sitting in a steaming sauna, or engaged in vigorous exercise. At the cellular level, cells are constantly dividing and reproducing, of course, processes that even in healthy people may frequently and hopefully transiently go out of control. If uncontrolled growth persists, it would lead to tumors and cancers, but a fully functioning immune systems will quickly restore growth to its normal levels. At the molecular level, regulation is similarly complex. As John Maddox puts it: "A cell is a self-regulating biochemical democracy in which the several parts are continually casting votes in the form of the chemical signals they transmit. The genome is to the cell as the Supreme Court is to the national judiciary" (Maddox 1998, 164).

Other types of feedback would seem to operate at much higher levels of organization such as communities, nations, or perhaps even civilization itself—their study generally going under the name of *cybernetics*.[4] Even though individual human behavior would seem almost infinitely variable,

---

4   Founded by the American mathematician Norbert Wiener in the 1940s to examine the role of various feedback mechanisms in such diverse areas as systems control, computer science, philosophy, the organization of society, and in biology itself.

such skilled operators as advertisers and politicians seem able to identify traits to which a surprising number of us conform and on which they can focus their manipulative powers. Depressingly, therefore, many of us can be persuaded to eat when we are not hungry, and to believe half-truths as gospels. However, the control that most interests me in this context is not necessarily the result of any purposeful action, and was first described by the Scottish philosopher and political economist Adam Smith. In 1776, he published his *Inquiry into the Nature and Causes of the Wealth of Nations*, in which he argued passionately for free trade. This was a very heretical view at the time as the conventional wisdom strongly held that the total volume of trade was fixed by the supply of gold and silver. In Book 4, Chapter 2, in what became one of the most quoted passages in economics, he wrote (the italics are mine):

> Every individual . . . neither intends to promote the public interest, nor knows how much he is promoting it. By preferring the support of domestic to that of foreign industry, he intends only his own security; and by directing that industry in such a manner as its produce may be of the greatest value, he intends only his own gain, and he is in this, as in many other cases, led by an *invisible hand* to promote an end which was no part of his intention.

Diversity of opinion is endemic in economics, but Smith's opus has been celebrated at every major anniversary since its publication. In 1976, for example, hundreds of assessments were published,[5] leaving no doubt that even after 200 years his work is still influential. There has also been a considerable debate on precisely what Smith meant by his invisible-hand reference (Grampp 2000, 443), his supposed reasoning including the forces arising from altruism, a joke, or merely luck. However, I would like to adapt Smith's term to give it a slightly different meaning. It may or may not be what he had in mind, but it seems consistent with his writings.

As I see it, Smith identified a powerful social feedback mechanism that promotes growth and prosperity. The operational details of Smith's invisible hand may not be understood, but by his reference to invisibility Smith implies that understanding is unnecessary; we should merely sit back and allow it to work. That begs the question, of course, but my interpretation of Smith's meaning is that individuals should be free to form their own judgments on what is important, and to do whatever they

---

5   See, for example, Terence Hutchison's article on Adam Smith's *The Wealth of Nations*, published on the book's bicentennial, which gives references to many other reviews published at earlier anniversaries (Hutchison 1976, 507).

believe is necessary to bring their ideas to fruition. In another of his books, *The Theory of Moral Sentiments*, he writes about altruism and a person's derivation of pleasure from another's happiness although, as he puts it, "he himself derives nothing from it" (Smith 1759)—which is perhaps an inverted schadenfreude. But that person may derive something from it. If one can see that happiness stems from one's actions, regardless of any personal benefit, that person would have had the pleasure of achievement. That in itself may be sufficient reward, especially if one could take personal pride in the social benefits flowing from what one had done.

Following Robert Solow's transformative discovery in economics (see Chapter 2), we now know that technical change is by far the dominant source of long-term economic growth, but such change can come in many forms. Its technological component is well understood, and global expenditure on the search for new and improved technologies is enormous. However, although the pursuit of efficiency has always been an institutional priority, it is only in the past few decades, following the revolutionary developments in computing and communications, that it has been possible rigorously to implement that pursuit. Consequently, institutions now revel in the powers of their new toys. We now live in an age in which efficiency— that is, *perceptions* of efficiency—is paramount, particularly in resource allocation and use. But the relationship between efficiency and creativity is not understood.

This serious situation should be of concern to everyone because as members of civilized society we are all stewards of creativity. Before the approximate watershed date of 1970—the "dawn of the age of efficiency"— we could safely assume that creativity, like the blooming of wildflowers, would take care of itself. Provided they were sufficiently determined, pioneers were generally free to tackle any problem that interested them. That assumption is now invalid. For technology, pioneers can usually get backing because they can point to tangible objectives whose potential benefits can be assessed. Unfortunately, science is dominated by philosophy. Peer endorsement for radical (but abstract) challenges is therefore seriously problematic simply because an advance statement of a justifying case cannot always be made.

It is likely, therefore, that a latter-day Adam Smith might have to deal with such criticism as: "Mr. Smith, you claim that although every individual intends only his own gain, he is led by an invisible hand to promote ends that were no part of his intention. *But where is the proof for your assertion?*" As the eighteenth-century Adam Smith had offered none, his radical views could have been safely dismissed as merely an expression of opinion. However, a sufficient number of influential people appreciated the value of his thinking, and so Smith's ideas took root. Rigorous proof might have been unnecessary because his assertion struck a chord with their experience and understanding of human behavior. That is, they were free to take it on faith without having to subject it to the interminable rounds of bureaucratic assessment our new age routinely demands. Today's world is truly bizarre. The emperor's new clothes are constantly admired although many can see that he is stark naked. But Smith's invisible hand and other philosophies advocating freedom seem to be denied *because* it is impossible to see them. They cannot be rigorously assessed, therefore.

My assertion is that the guiding force behind Smith's invisible hand is *creativity*. When most, if not all, researchers were free to explore, Smith's invisible hand could work its magic. Growth was fostered, therefore, even though it was not part of scientists' original intentions. Indeed, as my philosophical wizard pointed out in **Poster 3**, that is precisely what happened up to the so-called Golden Age of economic growth that began to end around 1970. As civilizations develop, populations expand. Thanks to Mr. Solow, we know that economic growth is led by technology but diminishing returns will soon follow unless we find new technological veins, and nowadays their most reliable source is new science. As things stand today, however, the funding agencies will allocate freedom only when they agree that researchers' objectives are appropriate to today's circumstances. As a result, putative members of a twenty-first-century Planck Club are highly likely to be frustrated. Thus, our proxies (politicians and other leaders, public and private research funding organizations, etc.) are seriously undermining a vital part of the feedback that prevents civilizations from becoming unstable and that keeps us away from the Damocles Zone.

Instabilities of one kind or another finally get us all eventually, of course, but experience shows that we do have some control over when it might happen. Life expectancy at birth in 2000, for example, was some 77 years in the United Kingdom and the United States, while it was some 37 years in Zambia.[6] In contrast, in 1842 in Manchester (UK), for example, a professional male had a life expectancy at birth of 38 years, whereas that of a manual worker was only 17 years (Wood 1991, 21). These improvements might cavalierly be attributed to the fruits of a developing civilization, but they stem entirely from ingenuity, especially in science and technology. There seems to be no reason in principle why these favorable trends should not continue, albeit perhaps at reduced rates, but institutions must understand that we can reap rich harvests only if scientists are free to cultivate their creativity.

---

6   US Census Bureau's International Database.

The figures on life expectancy alone indicate that although there is no escape from the fact that complex systems on average always evolve toward instability, evolution rates of specific systems should be controllable if we apply the appropriate feedback. But first we must find them, of course. Initially they may be invisible, and so our proxies must maintain the environments in which creativity can flourish as they did until relatively recently. For any given system, therefore, progression toward the Damocles Zone is not necessarily inevitable. Indeed, it would seem that we might defer entry indefinitely on any human timescale provided that the warning signs are recognized and appropriate actions taken.

What might those signs be? Jared Diamond has described the events that supposedly led to the Easter Island collapse, but as I mentioned, those who criticized the crazy policy of building escalating numbers of stone statues would probably have had no forum. If they had, collapse might have been avoidable. The appropriate actions in that case would have been for the islanders, when they took up residence on the island, to invite a few of their number to advise on the implications of policy decisions—in today's language, they might represent, say, an Ecology Research Unit. Had they done so, it seems likely that this unit would have noticed the signs of deterioration as the island moved inexorably toward the Damocles Zone.

It might have worked something like this. As the island is normally swept by very strong winds, one might reasonably expect the Unit's scientists to monitor tree movement as a function of wind speed at treetop level, and to measure the corresponding wind speeds at ground level. As the islanders continued to cut down trees more rapidly than new ones could grow, the scientists would probably have noticed that as tree cover was reduced some tree movements sometimes approached their elastic limits even when winds were unexceptional, and that wind speeds at ground level were also increasing. Thus, it should have been obvious that the island's ecology was beginning to lose the flexibility to withstand elements it had successfully resisted for millennia. A reasonable reaction in those circumstances would have been for the Unit to seek, say, a moratorium on statue building for a few years. They could have told the authorities that if they continued to cut down trees, then one day, probably without warning, they would not only have to manage with fewer trees but with none at all, with all the grave consequences that would entail. Had the authorities heeded my imaginary Ecology Research Unit's warnings, they might then have agreed on a reasonable rate of stone-statue building, entry into the Damocles Zone might have been avoided, and Jared Diamond might have written about a sustainable success rather than a catastrophic collapse.

The world is vastly more complicated than tiny Easter Island, of course, but many great and extended civilizations have either collapsed or dwindled in the past. We may not fully understand the precise reasons for their passing, but a few causes arising from various forms of mismanagement seem dominant. Thus, Mesopotamia's poor irrigation policies led to increasing soil contamination and starvation—ancient Rome's corrupt bureaucracy had ambitions outstripping the capacity of its stagnating economy. Today, it is not difficult to see increasing causes for concern. As a would-be member of a putative Global Prosperity Research Unit, a group that would assess and advise on *all factors* that might influence global prosperity—intellectual as well as material—it seems that per capita economic growth is an important indicator. Material prosperity is not everything, of course, but it does seem to be a necessary if not sufficient condition for global prosperity. The rates of growth in world gross domestic product (GDP) seem to be hovering around 1.5% per person per annum. If that figure is correct, would it be adequate? Would the scientists at my imagined Global Unit be happy that a 1.5% margin above stagnation will keep us clear of the Damocles Zone?

Another cause for concern is that the world's economic systems are becoming increasingly monolithic. This tendency might be stimulated by the emerging economies' wish to increase their share of global prosperity. *The Economist*, in a "survey of the world economy" published on September 16, 2006, said:

> Last year the combined output of emerging economies reached an important milestone: it accounted for more than half of total world GDP (measured at purchasing-power parity). This means that the rich countries no longer dominate the global economy. The developing countries also have a far greater influence on the performance of the rich economies than is generally realised. Emerging economies are driving global growth and having a big impact on developed countries' inflation, interest rates, wages and profits. (Woodall 2006)

Thus, for example, in 2006, China became the largest holder of foreign exchange reserves,[7] and the British steel industry was transferred to the ownership of Indian capitalists. The energy and vigor of the emerging economies should indeed increase global prosperity, but surely their effects will have a sustained impact on growth only if their contributions are

---

7   Holding $941 billion, China narrowly overtook Japan in 2006. Taiwan was the third largest holder, with Russia fourth. See *The Economist* p. 98 (Sept. 2, 2006).

new, and not merely based on producing existing ranges of goods and services more efficiently and cheaper than the advanced world can.

However, it should go without saying that monolithic systems lack diversity, that traditional font of ingenuity. By their very nature, such systems seek to impose uniform structures of customs and practices, and by promoting harmonization they discourage individuality. Thus, individuals or nations must either conform or be prepared to face the pressures arising from nonconformity. This trend does not seem consistent with the enhancement of global stability.

Economic growth may not be everything, but buoyant growth creates optimistic environments and the resources to deal with the trials and tribulations that our proxies should expect to beset us from time to time. Today, it seems generally agreed that the most important problems facing humanity include terrorism, the rise in religious fundamentalism, pollution and global warming, poverty and disease, the security and availability of energy supplies, and the huge potential increase in the resource needs of the emerging economies. China and India together, for example, represent approximately 40% of the world population. It is an awesome list, especially as various pundits at various times have described *each one of these problems* as representing the gravest threat to global stability. Thus, humanity seems to be faced with a diversity of slippery slopes leading to Damocles Zones regardless of whether my own concerns are included.

The list of problems may be daunting, but they may not be the most important. In 1957, a group of senior California Institute of Technology (Caltech) scientists ambitiously published their forecasts for the next hundred years in a book (Brown et al. 1957),[8] based on deliberations at some 30 conferences involving senior industrialists and other leaders. Their estimate of the most important problems the next hundred years might bring included the threat of nuclear war, population growth and food production, resource allocation, and the problems arising from the spread and intensification of the advanced world's (particularly the United States') industrial culture.

The book contains fascinating discussions on humanity's prospects and problems from the perspectives of 1957. It was, of course, a nightmare time when miscalculation could with very little warning have plunged the world into a devastating nuclear war. That threat may now have subsided,

---

8   This august publication includes a foreword and a postscript written by Sir Solly Zuckerman, who was later to be the UK's first chief scientific adviser, and a preface from Lee A. DuBridge, president of Caltech and a senior colleague of Vannevar Bush (see Chapter 3) during World War II with overall responsibility for the development of radar. I am grateful to Terry Clark for drawing this book to my attention.

but the authors very appropriately drew attention to the perils of forecasting, as they should. Even though the book was written by eminent scientists, it contains no inkling of the dramatic revolutions in electronics and communications that began to pervade the world only some 20 years later. As for their fears for world food production, the authors would also no doubt have been astonished to learn about the Green Revolution, which began to transform agriculture in the 1960s. Within a few decades of their forecasts, some parts of the world were plagued by so-called food mountains, and their governments paid farmers *not to produce food*. However, it would be most unwise to assume that such profligate policies will endure. In the light of the increasing demands from the developing countries, it would not be surprising to see food production once more on the growing agenda of critical problems facing humanity.

Although none of the problems that Brown et al. identified have entirely gone away, one can see the marked change in emphasis. Their book concludes: "The problems which we face in the years ahead are indeed both numerous and grave, but, theoretically at least, it seems likely that they can be solved by the proper application of our intelligence" (Brown et al. 1957, 152).

For the few decades following the book's publication, authorities did indeed ensure that intelligence was properly applied. Notwithstanding, therefore, that the authors assessed the problems as "numerous and grave," they did not, in sharp contrast with today's world, recommend that humanity's intellectual resources should be marshaled into deriving specific solutions to each problem. The academic authorities continued to allow intellectual endeavor full and free rein because that's what they had always done. Significant proportions of industry did much the same. It does not follow, of course, that we can automatically attribute the high and unprecedented rates of economic growth of the Golden Age to those simple policies. There may be other reasons. But, as a follower of Solow, my working hypothesis throughout this book has been that unconstrained creativity eventually leads to new opportunities and new growth. Conversely, although directed creativity may sometimes be advantageous in the short term, it eventually leads to diminishing returns and falling growth. As the authorities' actions over the last 25 years or so seem to have had precisely that effect, while it is not rigorous proof—always problematic in economics—the hypothesis would seem to merit serious attention.

Today's problems might not prove as transient as some of those foreseen in 1957. To make matters worse, the fact that the world is intrinsically nonlinear means that the spontaneous creation of new nonlinearities cannot be ruled out. Furthermore, each one will need to be controlled by its own feedback mechanism if stability is to be maintained. Thus do the events of history progress. Even the most carefully prepared predictions will probably be wrong, therefore, and our supposed list of current problems may still be far from complete. However, this is not a prescription for despair. Our best strategy in these circumstances should be the old one of striving to understand as much as possible about the present in the reasonable hope that it will be sufficient preparation for the future. We are not straws in the wind. Our very progression from the ranks of the primitive primates 5 million years ago seems entirely due to the random flowering of our innate intelligence. It has seen us through ice ages, plagues, wars, floods, droughts, and other environmental mayhem. There is every reason to expect that its "proper application" will continue to see us through indefinitely.

As things stand at present, that optimism hinges crucially on that simple word "proper." There is no doubt that the potential power of intellect is widely appreciated—the pen is mightier than the sword and so on— but the nature of its chief characteristics seems not to be understood at all. History shows that the more powerless creative individuals become, the more they are immersed in environments that institutionalize dogmatism. Very few new ideas emerged from the suffocating environment imposed by religious dogma during the Dark Ages, for example. Creativity is a delicate plant. Everyone who has had an idea is usually plagued by doubt and uncertainty. Is it really original? Is it correct or valid? Does it matter? For creativity to flourish, it does not necessarily need encouragement. Indeed, since its origins are not understood it may be impossible to encourage. But intellectual pioneers need environments that *accommodate* dissent, as I tried to explain in my *Pioneering Research*. Should it be surprising, therefore, that many seem to have lost their inspiration when they must struggle every day with the all-pervasive dogma on efficiency and accountability, however well intentioned its originators might be?

Pundits may presume the current threats the greatest ever, but the ill-considered actions of our proxies are making them much worse because they are undermining the creation of the very feedback mechanisms that have always kept us from the brink. Our universities have been reservoirs of creativity for the past 900 years. They have served us superbly well, as long as they have been free.

Industry, too, has a crucial role. Unfortunately, the great companies seem to have virtually ended their support for exploratory research. Today, it has been deemed that technology rules. Research is now the servant of technology, and apparently, each project must prove in advance that it can pay its way. In May 2004, the British Petroleum (BP) Group's vice president of technology said in a speech entitled "Technology: Demonstrating Value to the Corporation": "Nevertheless technology, like every aspect of what goes on in any well-run business, constantly has to justify itself—to demonstrate value to the corporation. It's not, and never can be, an end in itself."

How times change! In the 1960s, IBM's chairman, Thomas Watson Sr., began the Fellows Program, in which he appointed Fellows (he called them his "wild ducks") for five years to be "dreamers, heretics, mavericks, gadflies, and geniuses." Their remit was simply to "shake up the system." The Fellows Program has been supremely successful. Only some 165 scientists were appointed, but five of these won Nobel Prizes. General Electric and Bell Laboratories ran similarly distinguished programs. In 1980, as I have mentioned, BP launched Venture Research, arguably one of the most ambitious and imaginative exploratory research initiatives in industrial history, and supported it for 10 years. But then came a recession. In 1992, IBM suffered the biggest loss in US corporate history, and the company was "reborn" shortly afterward. But common sense prevailed, and the company still runs the Fellows Program, although its Fellows now seem to have somewhat less freedom than wild ducks typically enjoy— among other things, they are now *expected* to advance IBM's technological leadership.

Nevertheless, IBM today seems to be one of a very few major companies that appreciate the full value of unconstrained intellectual endeavor. Nicholas Donofrio, IBM's senior vice president, technology and manufacturing, said in his 2004 Hinton Lecture:

> About 25% of what we spend in research we spend on what we would call pure research. It may be maths—I am sure you remember Benoit Mandelbrot fractals. I am not sure that they sell computers, by the way. . . .
>
> . . . We did all this work with the scanning tunneling microscope, not knowing what would come of it. We did the basic work on high-temperature superconducting materials that has at least prodded some other people to do even more seminal work in that area. We will never capitalize on that work to be candid with you, but that does not bother us, because smart people like to be near smart people. We have a very simple philosophy. If you have one or two Nobel Laureates, I do not care whether they are working in that area or not. If they go to the cafeteria and people say, "I saw her," or "I saw him," it is great. . . . (Donofrio 2005, 24)

These prescient remarks are extraordinarily courageous for an industrial leader in today's climate, and reminiscent of past industrial visionaries. Even senior academics might think twice about advocating expenditure that might be perceived as leading to less-than-optimal returns on investments of "tax dollars" or other currencies for fear of being trumped in their bids for funds.

The pharmaceutical companies—known colloquially as *pharmas*—have traditionally made considerable investments in research. Their research budgets have increased some fifty-fold since 1970, and nowadays big companies might devote more than $5 billion a year to R&D. Yet there is widespread concern about the decreasing output of new drugs. As ever, the discovery of new products depends on having research environments that encourage flair and creativity. Instead, as Pedro Cuatrecasas points out: "Scientists must contend with 'management by objectives,' hierarchical and autocratic organisations, mandates from strategic planning groups, detailed and rigid scheduling, constant reporting, and achievements driven by milestones and flowcharts" (Cuatrecasas 2006, 2837).[9]

The role of product champions, so essential in industry, has virtually disappeared as the consequences of being "wrong" in today's climate can be severe. In the sciences generally, as I have explained, twentieth-century Planck Club members would be unlikely to get funded today. Similarly, virtually every "blockbuster" drug ever marketed (AZT, acyclovir, cimetidine, fluoxetine, etc.) would be unlikely to survive what Cuatrecasas describes as the current well-managed and efficient go/no-go systems. Moreover, companies are increasingly reviewing their activities against what others are doing (benchmarking), rather than exploiting their own skills and experience.

Never before, therefore, have we been in greater need of people in any walk of life who will "shake up the system" and liberate it from second-guessing bureaucracy. With the possible exception of global warming (see **Poster 4**), most of our current problems stem from human actions or neglect, and in principle, therefore, there is no reason why we should not solve them. On the basis of past experience, directed solutions—for example, the development of radar during World War II—are possible only if the intellectual environment is sufficiently fertile to give the authorities these options. Otherwise, they will be little more than a waste of money. Finance is crucial, of course, but that should not be a problem if economies are buoyant, as they are likely to be if their scientists are free. Thus, creativity

9   I am grateful to Desmond Fitzgerald for drawing Cuatrecasas' paper to my attention.

is at the heart of a powerful positive-feedback loop. The authorities seem unaware of that simple fact; indeed, they are acting to inhibit it.

In summary, therefore, essential steps for avoiding collapse would seem to include the following:

- The establishment of research initiatives aimed at creating a twenty-first-century Planck Club. In time, we could have a global network of such initiatives that should eventually lead to increased economic growth and buoyant economies.
- The emergence of altruistic sponsors to help fund these initiatives.
- The creation (or perhaps the re-creation) of an extensive network of universities that will encourage and foster scientific freedom.
- The emergence of industrialists who will convince shareholders that a small proportion of industrial activities should be free of short-term assessment.

I will suggest how they might be taken in the following chapters.

2   Scientific Freedom and Transformative Research

The Foundation must support the most innovative and potentially transformative research—research that has the capacity to revolutionize existing fields, create new subfields, cause paradigm shifts.... Ongoing review and evaluation make clear to the Board that the Foundation's current solicitation, review and selection processes must evolve, in some respects substantially, in order to achieve the transformative potential the Board envisions. NSF will...create an environment that is more open to and encouraging of such proposals from the research community.

—US National Science Board:
2020 Vision for the National Science Foundation, 2005, p. 7

There is hardly a researcher who does not hope, perhaps secretly, to make the observation that might lead to substantial progress or a major new discovery. No one can say in advance what form it might take, of course. The departure may be minor, but the expectant researcher may hope that it is the tip of an iceberg and an indication of a new facet of Nature's behavior. Whatever it is, that Eureka! moment so beloved of the media would be a most unlikely occurrence. In some respects, major discoveries can be like winning the lottery, but there is a very important difference. Lotteries usually notify the biggest winners. In research, however, it might not immediately be obvious that you are onto something new. Experiments are, in effect, dialogs with Nature, but what language is she using? Are you listening properly? Dudley Herschbach from Harvard University told me a nice story about this. When he won the Nobel Prize for Chemistry, he had many letters of congratulation from friends around the world. Although not a linguist, he could more or less decipher all except one. "What language do you think it was written in?" he asked. The anticipated responses of Chinese, Russian, Arabic, or Morse code were flicked away with a smile. "Braille," he said. "How do we know that Nature might not speak to us in the equivalent of Braille or something more esoteric?"

The great microbiologist Louis Pasteur (1822–1895) once said: "In the field of observation, fortune favors the prepared mind." Dedicated researchers should therefore be alert to the possibility that unusual results might mean that Nature is trying to tell them something. There will be no fanfare of trumpets, and unless they take positive action, the moment may soon pass. If you are lucky, you might then experience a sense of anticipation or elation, a sense that here is an opportunity that should be grasped if only you can work out exactly what it is. You will have to be careful, however, because it may not be good news. You may have misread the signs, or the unexpected result may indicate a failure in your equipment. You won't know in those first anxious moments, of course, and may be faced with days or months of troubleshooting until you can be sure that everything is working as it should. If you can eliminate the hardware, your problems could then become more serious. The design of the experiment may be at fault, and the surprising result may indicate that you are not actually measuring precisely what you set out to measure. Even so, you could still be on to a winner.

In 1963 Arno Penzias and Robert Wilson began their independent professional careers at the Bell Laboratories in New Jersey. They had both taken their PhDs in the previous year; Penzias from Columbia University, and Wilson from the California Institute of Technology. Hired to work on radio astronomy and satellite communications, the two young physicists seized the opportunity to use a huge microwave receiver—a 20-foot horn reflector antenna—recently made redundant from Bell's satellite telecommunications program. They planned to use it to study radio emissions from the Milky Way, but quickly ran into problems because all they could see was noise—an apparently uniform random radio hiss from every direction they looked. Nowadays, such an observation could be terminal. Commercial pressures do not allow open-ended commitments to solve what business might see as peripheral problems, especially if young recruits are involved. However, their employers did not insist that they give up, nor did Penzias and Wilson want to, and they began a systematic search for possible sources.

The story is well known (Wilson 1992), particularly their struggles with a family of pigeons nesting in the horn whose droppings seemed to be fouling their results. The point I want to make here is that after searching for possible faults in their system for about a year, a tour de force in commitment in itself, they concluded that the noise did contain a real signal but unfortunately it just happened to look like noise. They knew their system intimately, of course. They could therefore calculate precisely how noisy it should be. As their signal was bigger, some of it could be coming from a real source.

In research, the old adage that if it looks like a duck, walks like a duck, and quacks like a duck, it *is* a duck can be very seductive. Unfortunately, the obvious conclusion can sometimes lead you up the garden path. You may be seeing the shadow of something else. To make Penzias and Wilson's problems even more baffling, the signal was not only constant in every direction where they looked but also had no seasonal variation, and so perhaps it was coming from outside the solar system. If it were real, however, what could the amorphous hiss mean? After discussions with many colleagues, and particularly with Robert Dicke at Princeton University, they concluded that it was from the afterglow of the Hot Big Bang, the name given to the theory that the universe was created in a single cataclysmic event some 13.6 billion years ago. Their amazing discovery, often described as perhaps the greatest of the century, won Penzias and Wilson a share in the Nobel Prize for Physics in 1978.

It also led the scientific community, informed by the mysterious processes that constitute consensus, to interpret the discovery as *confirmation* of the Big Bang model. Indeed, the seeds of consensus germinated almost immediately following publication of their paper in the *Astrophysical Journal* (Penzias and Wilson 1965). This was probably because the adjacent paper was from Dicke's group "explaining" their result (Dicke et al. 1965). Unfortunately, these papers effectively blocked the search for alternative theories despite the fact that many researchers had published temperature estimates of interstellar space made independently of the Big Bang model that were also in broad agreement with Penzias and Wilson's measurement. There were alternatives, therefore. Nevertheless, the Big Bang model went from strength to strength. References in the literature today imply *that it is established fact*, a tablet of stone. One often sees such statements as:

- "The universe was created 13.6 billion years ago."
- "The galaxy was formed 1 billion years after the Big Bang."
- ". . . processes similar to those observed in the early universe . . . ."

But the model's status as *hypothesis* is rarely mentioned. This is a great pity, and the episode nicely illustrates the social pressures that can dominate science.

Leaving aside the history of its interpretation, would it be correct to describe the Penzias–Wilson *research* as transformative? I do not think it would. One might as well ask whether Planck's or Einstein's original work were *relevant* in today's language, a question that would hardly have occurred to them. They simply did what they thought was important. There were no other considerations. In the Penzias–Wilson case, they set out to explore radio emissions from the Milky Way. The field of radio astronomy was growing rapidly in 1963 although it was still in its infancy. Bell Labs had played a pioneering role 30 years earlier when they asked Karl Jansky to explore sources of static that might affect their transatlantic radio-telephone service. He concluded that static would not be a serious problem, but he went on to become the first to observe radio signals from the Milky Way. The finding attracted a lot of interest, and was published by the *New York Times* on May 5, 1933.[1] Unfortunately, it was not an interest that Bell Labs could sustain in a severe depression, and Jansky was obliged to move on to other things.

But Jansky's discovery was not forgotten. The Jodrell Bank radio-telescope in Cheshire, England was built in 1957. In 1963 the British government appointed a Radio Astronomy Planning Committee under the chairmanship of Lord Fleck (Saward 1984, 268). In another indication of how much customs and practices have changed in only half a century, Alec Fleck was not an astronomer but a distinguished industrial chemist (he was then chairman of Imperial Chemical Industries), although his committee included many astronomers. The Fleck committee made recommendations for work in five fields:

- The solar system.
- The galaxy and individual nonthermal sources associated with supernovas and radio studies of certain types of stars.
- Nearby normal galaxies.
- The radio galaxies that exist with considerably greater intensity than our own galaxy.
- Cosmology.

Thus, Penzias and Wilson's work could have fitted into the Fleck plan had it been operational in the United States. In today's jargon, radio astronomy was therefore widely recognized as a priority area when they began their work. It would have been relevant, therefore, and there would have been no need to set up a special initiative to foster it. There is no question that their *discovery* was transformative. However, one of the characteristics of research is that it should always offer the prospect of a surprising outcome. Any research has the *capacity* to be transformative, therefore, particularly for researchers armed with Pasteur's insight as Penzias and Wilson clearly were. But they must be allowed to keep at it for as long as it takes.

---

1   The *New York Times* headlines that day were: "New Radio Waves Traced to Centre of the Milky Way; Mysterious Static, Reported by K. G. Jansky, Held to Differ from Cosmic Ray. Direction Is Unchanging; Recorded and Tested for More Than Year to Identify It as from Earth's Galaxy."

How should transformative research be defined, therefore? I have been preoccupied with this question since 1980, when BP invited me to set up an innovative research initiative under the name *venture research*, and effectively gave me and my colleagues complete freedom in what we could do.[2] My preferred definition of *transformative research* (TR) would be the same as that for venture research.[3] Indeed, the terms are synonyms. My definition is:

> Research that sets out radically to change the way we think about an important subject.

That may be a necessary definition but it is not sufficient, and for the rest of this chapter I will try to put more flesh on these bones.

Why should a particular definition have special value? In principle, researchers should be free to do what they wish, as indeed was often the case before the 1970s or so. Scientists have not changed, but the environment in which they operate has. As I shall discuss in the next chapter, researchers today are rarely free to be full-fledged *scientists*—that is, to survey Nature's vast unexplored domains and to plan their forays according to their personal inclinations. Instead, they are subject to financial and intellectual pressures to deliver the results that consensus deems would be the best value for money. Thus, a generation of researchers has been encouraged to think parochially rather than globally, and not surprisingly, this has led to a dearth of major new discoveries. We hoped that our short definition, which was as succinct as we could make it, would also allow would-be pioneers all the freedom they needed.

When working on our strategy, we came to the conclusion that current procedures for selecting and funding research would have to be abandoned because they did not pass the Planck test. Our job, therefore, would be to identify the procedures Nature herself might use if she were obliged to operate *in today's circumstances*. It is perhaps the ultimate arrogance to claim such an affinity with Nature, but we did not present it as a take-it-or-leave-it position and specifically asked applicants for help; that is, we explained what we were trying to do, and invited applicants to set out their proposals in ways that Nature might respect. If they (or indeed anyone)

---

2   The Venture Research Unit drew its staff from mainstream BP. I was its head throughout the 10 years of its existence. The other scientists, who were members for two to five years, were Derek Barker, Peter Beadle, Keith Cowie, David Ferguson, David Ray, Tony Regnier, Tim Sanderson, and Jean Shennan.

3   The term transformative research perhaps speaks for itself, but venture research seems more exciting. Unfortunately, the term venture has been hijacked by the venture capitalists, and is usually taken to indicate a risky but hopefully profitable initiative within a specific field and timeframe. As I shall explain, the Venture Research Unit was not at all risky.

could argue that our procedures imposed artificial or contrived constraints, then we would immediately do our best to remove them.[4] Our selection strategy steadily evolved, therefore, always converging, we hoped, toward being as objective as possible. Indeed, our aim was to make the problem of research selection a rigorous discipline in its own right.

My experience spans 10 years operating the new procedures and a further 17 in which they have been further refined. As far as I know, no other scientist has more experience of TR. That does not give it special value, of course. I present it here for assessment in the hope that it may help prospective transformational researchers. Unfortunately, there is no forum for evaluating the discipline of research selection. Indeed, few funding organizations would accept the need for one. Instead, they expect reviewers and selectors to pick up the necessary expertise even though there is no agreed-to store of knowledge from which they can draw. Imagine the chaos that would ensue in physics, say, if funding organizations required reviewers to assess proposals on the assumption that there was no established learning! The work I describe in Chapter 7 has never been independently evaluated, therefore. I hope it speaks for itself.

So, how might it work for TR? Perhaps the most important decision facing prospective transformative researchers concerns their degree of commitment. If they are not intending to devote themselves totally to their new research, then it is unlikely to be transformative. As I see it, a wish to take part in a TR initiative implies a researcher's intent, subconscious or otherwise, to radically change how the world is perceived. It might involve challenges to axioms hitherto accepted as statements of fact, or the exploration of one of Nature's unfashionable domains, or a possible major departure from her expected behavior. Whatever the intent, however, it will most definitely need a determination to stick to one's guns.

Planck Club members might not necessarily have exhibited these characteristics when they were first setting out. In principle, they had probably been free to tackle anything, so it is possible that decisions to embark on ambitious crusades might not initially have been deliberate. They might even have drifted into them before their insight crystallized and their commitment became total. Their modern successors must be similarly committed, of course. Researchers with a part-time interest in a problem who say (or imply) that they want to change the world cannot expect to be taken seriously. However, researchers today do not have the luxury of allowing their ideas to germinate at their natural rate, and are

---

4 Unfortunately, some remained, but at least we were aware of them; we were able to offer support for only three years, for example, but support could be renewed for additional three-year periods. Indeed, it usually was, as can be seen in Chapter 7.

expected to deliver fully hatched proposals at the outset. As I shall discuss in Chapter 4, operators of a TR initiative must take these factors into account, and do what they can to encourage and stimulate prospective transformative researchers while their ideas are still at the embryonic stage.

We might do worse, therefore, than to define TR as the type of research done by the founder members of the Planck Club. The environment in which their prospective successors would operate is also important. It must foster scientific freedom, of course, and the first step in that direction concerns the funding agency's commitment. Pioneers setting out on their daunting journeys will want to be reasonably sure that funding would continue if they met the sort of difficulties that Penzias and Wilson had to deal with. In essence, that means that the agency should normally back the researchers' judgment. That does not mean that funds should be extended indefinitely, but it does imply unusual levels of partnership and trust between sponsors and researchers so that decisions on termination are not taken precipitously.

However, the TR specification should be restricted to the general and the abstract. Indeed, for Venture Research, which I hope might be seen as a precursor, the realization that we should not be precise in defining what we were looking for was one of the most surprising. *Precision* means that we would fall into the traps created when agendas are set externally. Such agendas abound, and their very existence has led to the need for research initiatives free from them. Unfortunately, our very open-mindedness also created a catch-22 for ourselves when we searched for alternative funds after BP pulled out. Had we told prospective patrons that we would concentrate on specific areas, we would have been inundated with proposals in those areas as the conventional agencies often are, and there would be nothing to differentiate us from them. On the other hand, possible patrons seem to lose interest when we do not tell them the specific objectives that their money would go toward achieving. It seems that helping to create a twenty-first-century Planck Club, whatever its members might do, is not sufficiently tempting. In my long-running quest to find replacement funds this would seem to have been *the* major reason why we have not yet succeeded.

If one should not be specific, perhaps a TR prospectus should give examples of possible areas of concern. There might be substantial benefits if researchers were encouraged to derive new approaches to old problems. I do not agree. I wrote the wizard's warning essay (**Poster 3**) with tongue in cheek and humor on my mind, but the message was a serious one. Creativity works best when it is free from constraints. If someone had said to Max Planck, "We don't understand thermodynamics. You're a clever chap, why don't you look at it?," I doubt if he would have spent 20 weeks on it rather than the 20 years that proved necessary, *because he would not have been working on his problem.*

One cannot be sure, of course. However, I doubt that anyone would have spoken to Planck in that way when he was starting out. Thermodynamics' shortcomings were not generally recognized—it took Planck's genius to do that. It will take similar genius to identify the serious weaknesses in current foundations and to discover how to fix them, but we have no idea what they might be.

Our experience indicates that researchers who wanted to substantially increase understanding in important areas where it is weak or nonexistent; who had developed carefully crafted, radical, open-ended, general questions that could come only from prolonged and perceptive consideration; and/or who had developed unique, convincingly argued and flexible lines of attack would, in effect, define what is meant by transformative research.

3   Mismanagement by Objectives: The Need for Fresh Approaches

The state of mind which furnishes the driving power here resembles that of the devotee or the lover. The long-sustained effort is not inspired by any set plan or purpose. Its inspiration arises from a hunger of the soul.

—Albert Einstein, in his Preface to Max Planck's book,
Where Is Science Going? Ox Bow Press, 1933, p. 13

For the first time in history, there is growing dissatisfaction that the research enterprise is not as productive as it used to be. It seems to be becoming staid and might be in need of a boost. In the United States, the Report to Congress by the House Committee on Science on September 24, 1998, *Unlocking Our Future, Toward a New National Science Policy*, said:

However, if limited funding and intense competition for grants causes researchers to seek funding only for "safe" research, the R&D enterprise as a whole will suffer. *Because innovation and creativity are essential to basic research, the federal government should consider allocating a certain fraction of these grant monies specifically for creative, groundbreaking research.* (75)

In the European Union, the European Commission's New and Emerging Science and Technology's Adventure initiative provided funding for ". . . visionary research across a very wide spectrum of science and technology and with an orientation toward the long-term."

However, applicants must satisfy two conditions: (1) that the research supported lies clearly outside the domains covered by the "thematic priorities" and (2) that the research supported is "high novelty, is highly ambitious, and has a high-risk/high-impact character." The expected impact of Adventure projects was judged in the first instance on scientific and technical capability, but this must be in areas where societal or economic benefits may be expected in the long term. The Commission's documentation (December 2004) on Adventure went on to discuss the type of projects *that would not be funded*. The list included (the italics are mine):

Research which is *open-ended*, without clearly-articulated tangible objectives in the form of a highly challenging scientific objective, or the creation of a basic technology. As an example, the objective of "increased understanding" alone would not be considered sufficiently tangible.

Max Planck must have turned in his grave.

One can hardly imagine such pronouncements appearing in the 1950s and 1960s, say, when the gloriously freewheeling research enterprise could still safely be relied on to produce a steady stream of radically new discoveries, much to the benefit of industrial activity and economic growth. In the 1950s, Robert M. Solow from the Massachusetts Institute of Technology began the work that established *technical change* as the most important source of economic growth (see **Poster 6**). He won the Nobel Prize for Economics in 1987 for this seminal work. As virtually every industrialized nation has introduced substantial changes to its arrangements for commissioning research and development (R&D) since 1970 or so, we should be able to see signs of those changes in global economic growth. Expenditure on R&D has never been higher—**Poster 7** gives some statistics. (See **Figure 1** for trends in US expenditures.) Growth data for the last half century are given in **Figure 2**. Each point on the curve is the average annual real-terms growth per capita for the preceding 10 years. Growth has steadily declined. As I explained in my *Pioneering Research*, the decline seems to have been caused by the very policies that were supposed to create a more efficient and productive research enterprise! I went on to suggest that this issue should be examined by professional economists, but apparently it continues to be ignored.

Poster 6
—

Economic Growth

Before the 1950s, the main sources of economic growth were thought to be
- Capital
- Labor
- Resources

However, Robert Solow, in a 1957 study of the US economy between 1909 and 1949, could not account for observed levels of growth using these sources alone, and found that some seven-eighths of it (in output per hour of work) was due to technical change (Solow 1992). His study has been subsequently refined by many workers and most notably by Edward Denison. His study of American growth between 1929 and 1982 concluded that education per worker accounts for 30% of the increase in output per worker and the advance of knowledge accounts for 64% (Denison 1985). Thus, as Solow pointed out in his Nobel Lecture, technology remains the dominant engine of growth, with human capital investment in second place.

Solow's work would therefore seem to have proved beyond reasonable doubt that the twin pillars of (1) scientific freedom, as the major source of new technology, and (2) university autonomy, as an important guardian of educational standards, do indeed play vital roles in maintaining global prosperity.

Growth seems to have accelerated over the past few years. However, another force that might stimulate growth has received an enormous and artificial boost recently, and may be causing the apparent upsurge. According to estimates published by *The Economist* on June 16, 2005:

> The total value of residential property in developed economies rose by more than $30 trillion over the past five years, to over $70 trillion, an increase equivalent to 100% of those countries' combined GDPs. Not only does this dwarf any previous house-price boom, it is larger than the global stockmarket bubble in the late 1990s (an increase over five years of 80% of GDP) or America's stockmarket bubble in the late 1920s (55% of GDP). In other words, it looks like the biggest bubble in history.

Will we be well placed when this mega-bubble bursts? It seems unlikely, a view shared by many politicians, industrialists, and senior academics. It seems agreed, therefore, that the status quo cannot continue. Global markets rally from time to time, but each rally invariably turns out to be a "dead cat bounce" in the modern jargon. The underlying trend in real per capita growth would seem to be spiraling downward, with dips in each successive cycle becoming dangerously deeper. The urgent need, therefore, is for those who influence the markets to rediscover Solow's seminal work and to press for initiatives that can break the downward trend.

Expenditures, Millions

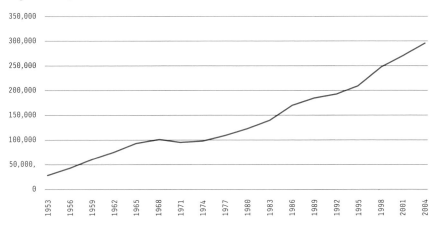

Figure 1: US real R&D expenditures (2000 dollars). (Source: US National Science Foundation, *Science and Engineering Indicators*, 2006.)

Poster 7
—

R&D Investments

The total R&D expenditure of all Organization for Economic Cooperation and Development (OECD) nations in 2005 was $687 billion in current money, or 2.3% of gross domestic product (GDP) (OECD MSTI Database 2005). This total includes the contributions from private industry.

In the United States the proportional expenditure on R&D was even higher, at 2.7% of GDP (see **Figure 1**), and has risen roughly linearly in real terms since 1953.

The average OECD gross domestic expenditure on R&D as a percentage of GDP was
- 1.95% in 1981
- 2.3% in 2005

The average annual global per capita real-terms economic growth was
- 2.5% for 1951–1976
- 1.5% for 1977–2004 (see **Figure 2**)

Percentage Change

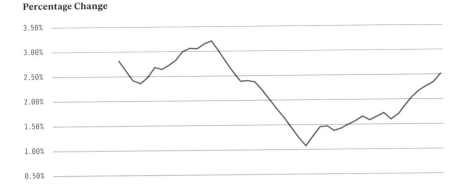

Figure 2: World real GDP per capita; rolling 10-year average percent changes. Each data point is the average for the preceding 10 years. (Source: Data for 1951–1992 have been taken from Angus Maddison, *Monitoring the World Economy 1820–1992*, OECD, Paris, 1995. These data have been extended by those from the International Monetary Fund for 1993–2004, the latest available upon initial publication. World population growth has been taken from the US Bureau of the Census, International Database.)

Initiatives are not always seen as good news. Researchers must be weary of the task of scanning the literature for details of the latest launchings that might possibly help in their never-ending quests for funds. All too often, they end in having to write yet more proposals that have only an approximately 20% chance of being funded. These colossal wastes of time and energy generally go unmentioned in the media. Why, therefore, should a transformative research (TR) initiative be of special interest? My answer is that the initiative should aim to bring about at least a partial restoration of academic freedom to its pre-1970 levels,[1] and therefore to correct perhaps the biggest imposition of the new policies.

All twentieth-century Planck Club members are now famous, of course, and many have acquired Olympian status. Researchers may conclude therefore that TR initiatives would be super-elitist and therefore of no interest to them. However, it should be remembered that every one of this auspicious Club's members started out as we all do, as humble searchers after truth. They did not follow some magically preordained path.

1   In 1940, the American Association of University Professors agreed on a Statement of Principles on Academic Freedom and Tenure, which was subsequently endorsed by over 100 professional organizations. It was reaffirmed in 1970 and in 1990. Regarding research, it says: "Teachers are entitled to full freedom in research and in the publication of the results, subject to the adequate performance of their other academic duties; but research for pecuniary return should be based upon an understanding with the authorities of the institution."

They, too, experienced doubt and uncertainty. What problem should I choose? Do I have sufficient ability to make progress? In another field of endeavor, Napoleon Bonaparte once said to his troops: "Every one of you carries a marshal's baton in his knapsack; it is up to you to bring it out." Napoleon could say that because the environment he had helped create encouraged such advancements. Scientists have always needed patronage, of course, but in principle, their progression thereafter depends only on their dedication and commitment. I have quoted Albert Einstein's wonderful remarks on Planck's motivation at the beginning of the chapter. Today, it is also most likely that researchers thinking deeply and driven by "a hunger of the soul" will at some time in their careers have a great idea. If it could be brought to fruition, it might turn out to be transformative and qualify the researcher for membership in a future Planck Club. One never knows, of course, when that might happen, but a TR initiative standing by to make their "labyrinthine paths" possible could transform their lives, and should be of interest, therefore, to *every* practicing scientist. A TR initiative would not of itself bring a return to the pre-1970s environment, but for would-be revolutionaries it could be the next best thing. In these circumstances, one might imagine my mythical Nature in extravagant mood saying to her troops: "Every one of you carries a ticket to Stockholm in your lab coat. It's up to you to find it."

Poster 8
—

An Even-Handed Universe? A Tale of Two Experiments

Throughout the history of science, it has been tacitly assumed that the universe is even-handed; that is, if Nature allows a process she also allows its mirror image. The entity that seems to be conserved here is called "parity," and put more scientifically, its conservation law says that all physical results should be independent of whether the observer uses a left- or a right-handed coordinate system. Confidence in this law was virtually absolute, and it therefore went unquestioned (apparently). However, in 1956, two theoretical physicists, Tsung Dao Lee and Chen Ning Yang, published a paper pointing out that for the weak interactions, that is, the forces mediating radioactive decay, there was no evidence either for or against the law, and suggested methods for testing it (Lee and Yang 1956). Within weeks, Chieng Shiung Wu and her collaborators carried out a very simple experiment that observed a violation (Wu et al. 1957). Lee and Yang were awarded the Nobel Prize for Physics in 1957 for their astounding discovery.

However, the law had been tested long before by Cox and his collaborators (Cox et al. 1928). They had also found evidence for a violation, but such was the weight of theoretical argument to the contrary that it was totally ignored even though the results of their experiments were published in a prestigious journal. As P. M. S. Blackett, a 1948 Nobel Laureate, said in his Rutherford Memorial Lecture of 1958:

> If these experiments had been followed up by any of us younger colleagues of Rutherford, many of whom were at the time looking for new and simple experiments to do—for this period at the Cavendish was less than usually active—then . . . the essential asymmetry of nature between left and right-handed systems would have been established nearly thirty years earlier than in fact it was. (Blackett 1959, 310)

There would seem to be nothing that Cox and colleagues could have done about this situation. They courageously set out to observe an effect that was most definitely not supposed to happen. But they did observe it, and their publication survived peer-review scrutiny (the old sort), which was a tribute to the reviewers. However, the work has rarely been cited presumably because their observation lacked a critical mass of scientists to endorse it. They had dramatically challenged established wisdom, but there was not yet a framework within which their contribution could be fully assessed. The flame of their truth was therefore quietly snuffed out, quenched not by conspiracy but by the indifference of the scientific community. Its time had yet to come, but Cox and colleagues would not get the credit they richly deserved.

Setting up a genuine TR initiative will not be easy, and the status accorded to peer review, or, more accurately, *peer preview*, will be crucial. There is, however, another serious problem: a researcher's transformative discovery may actually be ignored. **Poster 8** gives one relatively unknown example, but there are many others. Linus Pauling had unconventional views on vitamin C; Albert von Szent-Györgyi had unconventional views on almost everything; Barbara McClintock's transformative ideas on jumping genes were ignored for years, as was Peter Mitchell's chemiosmotic theory. All of these latter scientists eventually won Nobel Prizes, but the essential point is that the individual scientist has no control over the scientific community's opinion. Indeed, it is probably uncontrollable. As I mentioned earlier, the routes to consensus are mysterious. A TR initiative can avoid it as far as research selection is concerned, but according to science's unwritten law, a discovery does not count until it has been repeated by third parties and accepted by the community in general. Consensus rules in that domain.

As I mentioned in the last chapter, we concluded that we could succeed as Nature's proxy only if we could find selection procedures that she might approve of, and peer review[2] would not be one of them because it did not pass the Planck test. This rigorous criterion should, of course, be the *acid* test for all new initiatives. However, even Venture Research was not entirely free of peer review's clutches. Such is its all-pervasive power that our BP board could not bring itself to forgo it, and we were obliged to find a form that would satisfy us all. The solution was to split the problem into two; we would first make up our own minds, but before we took our recommendation to the board we would approach a single peer selected by the researcher as being the most important person likely to be supportive. We then fully briefed that person, both verbally and in writing, and did what we could to persuade him or her to endorse our decision. However, even this farcical, vestigial form of peer review often failed to produce the support we requested. In every case, we were able to persuade the board to go ahead anyway. This is a luxury few researchers usually have. Their proposals must speak for themselves. On the other hand, Venture Researchers had an active friend at court, and their subsequent successes proved that we were right.

To some extent, the authorities' reverence for peer review is understandable because it works reasonably well in the mainstreams. Since these make up the overwhelming majority of all research, this would seem to be an excellent record. Critics have a serious problem, therefore: *Peer review is so infernally convenient.*

But the general reader might appreciate a few clarifying remarks here. Peer review works best when its ministrations are confined to well-established disciplines. Each discipline can be regarded as a separate country with its own language, and peer review in that metaphor is language-specific. Funding agencies make special arrangements for the relatively few multidisciplinary programs they launch by recruiting linguists

---

2   Lest the reader should think that my views on peer review are extreme, Louis Alvarez, winner of the Nobel Prize for Physics in 1968, said in his autobiography: "In my considered opinion the peer review system, in which the proposals rather than the proposers are reviewed, is the greatest disaster to be visited upon the scientific community in this century" (Alvarez 1987, 200).

from each specialty. However, the agencies do not provide for *polyglot* peer review, or ways in which one's ideas can be assessed in a universal language like Esperanto, say. Put less fancifully, there is no peer review for science as a whole, which is, of course, the only language that Nature speaks. I do not know of a funding agency that has yet faced up to this inconvenient truth, or that even acknowledges its existence. Peer review, therefore, like the admission arrangements for an exclusive club, creates powerful barriers[3] for new entrants to a field. But genuine explorers may not speak the local language. They may even want to change it!

As every researcher knows, Nature does not respect consensus. We cannot expect to actually make progress simply because we have agreed among ourselves that progress lies in a particular direction. Unless, of course, that direction is a simple extrapolation from what has gone before, but such objectives usually serve only to consolidate—unless researchers are free to follow interesting observations wherever they may lead, as Penzias and Wilson were. However, as Planck pointed out, we should not expect our colleagues to accept significant criticism of their favorite dogma until we have demonstrated either its flaws or the viability of the alternatives. (Acceptance may still be deferred, of course.) If there is little or no slack in the system, that may not be possible. Thus, it is irrelevant that selection by peer review works reasonably well most of the time. So might selection by lottery or by tossing a coin. The fact remains that if it cannot pass the Planck test, *peer review is fundamentally flawed.*

Thomas Kuhn, writing about "normal" research, his special word for research in the mainstreams, gave the following words of warning:

> Normal research, which *is* cumulative, owes its success to the ability of scientists regularly to select problems that can be solved with conceptual and instrumental techniques close to those already in existence. (That is why an excessive concern with useful problems, regardless of their relation to existing knowledge and technique, can so easily inhibit scientific development.) (Kuhn 1970, 96)

He wrote that in 1970, that is, just as the governance of science was about to be drastically changed. What a pity that the authorities ignored him.

---

3   These barriers apply to the most senior scientists. Dudley Herschbach (Harvard University) won the Nobel Prize for Chemistry in 1986. He contacted Venture Research a few days after the announcement with a proposal that would in effect require him to work as a theoretical physicist. He had tried the usual agencies, but they had repeatedly declined to fund him at the required level. Their response seemed to be: "Stick to your knitting." We wholeheartedly supported him. See Chapter 7.

We have now had some 30 years of the new arrangements, and so it should be possible to make at least a preliminary assessment of their effects on scientific discovery. But first, I should choose a benchmark. In **Table 1**, I give a very brief description of some of the advances made by 20 Planck Club scientists. They either completed their work before 1970 or so, or generally enjoyed pre-1970s freedom. It is merely a sample. The twentieth-century harvest was so rich that other lists could be produced according to preference. However, the work of these scientists alone transformed our lives. Indeed, I find it impossible to conceive of modern life without their contributions. **Table 1** associates a single set of names with each discovery. Many others subsequently made substantial contributions in each field, of course, but I guess that few would dispute my specific attributions. Each name resonates in my imagination, and no doubt in many others, too.

Table 1: Some Twentieth-Century Discoveries

Name	Success
Max Planck	Discovered that energy is quantized
Ernest Rutherford	Founded nuclear physics
Albert Einstein	Photoelectric effect; special and general relativity
Paul A. M. Dirac	Predicted existence of positrons
Wolfgang Pauli	Exclusion principle; predicted existence of neutrinos
Erwin Schrödinger	Founded wave mechanics
Werner K. Heisenberg	Founded quantum mechanics; uncertainty principle
Alexander Fleming	Discovered penicillin
Enrico Fermi	Built first nuclear reactor
Oswald T. Avery	Discovered that DNA is the genetic molecule
Linus Pauling	Seminal work on the nature of the chemical bond
Dorothy C. Hodgkin	Pioneered X-ray diffraction techniques
Max Perutz	Discovered structure of hemoglobin
Francis Crick and James D. Watson	Discovered double-helix structure of DNA
John Bardeen, Walter H. Brattain, and William B. Shockley	Invented the transistor
Dennis Gabor	Invented holography
Charles H. Townes	Invented the maser
Barbara McClintock	Discovered transposons
James Black	Discovered how to design targeted pharmaceutical drugs
Sydney Brenner	Pioneered molecular biology

Science, one of the world's most authoritative journals, has in recent years been publishing its year-end nomination for "Breakthrough of the Year"—described as the most significant development in scientific research in terms of its consequences for the advancement of science and for society. *Science's* editors produce the nominations, and also list nine runners-up. Complete listings for the last six years are given in **Tables 2–7**, with the Breakthrough of the Year listed as item 1 in each table. As six years is a rather short period over which to assess performance, I have extended these data over the previous 12 years with details of the winning Breakthrough of the Year only in **Table 8**. Before 1996, *Science* called these awards "Molecule of the Year," and these are given from 1989, when the award was devised, until 1995. The short descriptions are mine, as they were for **Table 1**.

Table 2: Breakthrough of the Year, and Nine Runners-Up: 2006

Source: *Science* 314, 1848–1855 (2006).

1. Proof of Poincaré's conjecture.
2. The sequencing of more than 1 million bases of Neanderthal DNA.
3. Observation of increasing ice-sheet loss in Greenland and Antarctica.
4. Discovery of a new tetrapod-like fish species.
5. Construction of a microwave cloak of invisibility.
6. Discovery of a new drug for the treatment of age-related macular degeneration.
7. Several discoveries of new pathways to speciation.
8. Discovery of techniques that allow proteins and cells to be studied beyond the diffraction limit.
9. Several discoveries indicating that long-term potentiation strengthens connections between neurons.
10. Discovery of a new small RNA molecule that shuts down gene expression.

Table 3: Breakthrough of the Year, and Nine Runners-Up: 2005

Source: *Science* 310, 1878–1885 (2005).

1. Evolution in action: understanding evolutionary mechanisms.
2. Planetary exploration by robots.
3. Molecular clues to the coloring exhibited by plants.
4. Neutron stars: new observations.
5. Mechanisms of brain disorders.
6. Isotopic differences between Earth and extraterrestrial rocks.
7. Molecular structure of a voltage-gated potassium channel.
8. Human sources of global warming.
9. Systems biology: cell signaling networks.
10. Fusion: approval for International Thermonuclear Experimental Reactor.

Table 4: Breakthrough of the Year, and Nine Runners-Up: 2004

Source: *Science* 306, 2010–2017 (2004).

1. Exploration of Mars: data from rovers.
2. Discovery of a new species of small hominid.
3. Evidence confirming possibility that primates can be cloned.
4. Condensate family tree extended.
5. Exploration of "junk" DNA.
6. Discovery of a pulsar pair.
7. Evidence for declining biodiversity.
8. Water: possible evidence of new structures.
9. Public health: new public-private partnerships announced.
10. Seawater: molecular techniques to study organisms at great depth.

Table 5: Breakthrough of the Year, and Nine Runners-Up: 2003

Source: *Science* 302, 2038–2045 (2003).

1. Dark energy dominates universe.
2. Identification of genes contributing to mental illness.
3. Global warming: planetary responses.
4. How small RNAs orchestrate a cell's behavior.
5. Nanotechnology: physicists and biologists identify new techniques.
6. Gamma-ray bursts: new observations.
7. Mouse embryonic stem cells.
8. Left-handed materials and anomalous refractive index.
9. Genetic sequencing of Y chromosome.
10. New anticancer drugs.

Table 6: Breakthrough of the Year, and Nine Runners-Up: 2002

Source: *Science* 298, 2296–2303 (2002).

1. Small RNAs operate many cellular controls.
2. Neutrino mass differences observed.
3. DNA sequencing: solving developing world's problems.
4. Detection of polarization effects in cosmic microwave background radiation.
5. Movies on attosecond scale (10^{-18} s).
6. Ion channels that respond to taste sensations.
7. Cryoelectron tomography: viewing structures of intact cells.
8. Adaptive optics.
9. How light resets the circadian clock.
10. Extending the hominid fossil record.

Table 7: Breakthrough of the Year, and Nine Runners-Up: 2001

Source: *Science* 294, 2442–2447 (2001).

1. Nanotechnology: the first molecule-scale circuits.
2. RNA interference: quelling cell activity.
3. Solar neutrino oscillations observed.
4. First publication of human and other genomes.
5. High-temperature superconductivity: new record.
6. How axons find their way to correct destinations.
7. IPCC[a]: most of global warming is due to increases in greenhouse gases.
8. FDA[b] approves new cancer drug.
9. Bose–Einstein condensates.
10. Carbon budget: resolution of recent differences.

a   Intergovernmental Panel on Climate Change.
b   US Food and Drug Administration.

Table 8: Science's Breakthrough of the Year, 1996–2000, and Molecule of the Year, 1989–1995

2000	An explosion of gene sequencing data in bacteria and humans.[c]
1999	Ability to isolate and maintain human pluripotent stem cells in culture.
1998	Evidence that the expansion of the universe is accelerating.
1997	Lamb cloned from a single cell of an adult sheep.
1996	Protease inhibitors and chemokines can block HIV[d] replication.
1995	Confirmation of a new state of matter: Bose–Einstein condensates.
1994	DNA repair enzyme system.
1993	Tumor suppressor protein.
1992	Nitric oxide.
1991	Buckyballs (buckminsterfullerenes) C_{60}.
1990	Manufacture of synthetic diamonds.
1989	Polymerase chain reaction.

c   According to *Science*, this was possibly the breakthrough of the decade, perhaps even of the century.
d   Human immunodeficiency virus.

Comparisons between these two sets of tables, that is, **Table 1** with **Tables 2–8**, are complicated by the fact that they cannot yet be viewed with equal degrees of hindsight. The recent discoveries may look quite different in, say, 30 years' time, when some of them might turn out to have been momentous. However, one factor seems to stand out from the awards of the last 18 years: they are mainly for the achievement of specific objectives, and many of them have a strong technological flavor. Judging by these awards, pure, disinterested research seems to have taken a back seat

Numbers, Thousands

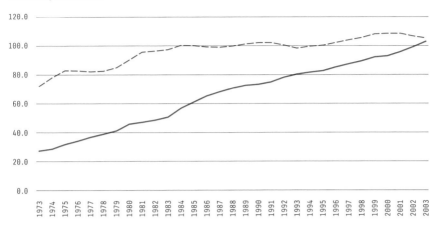

Figure 3: Science and engineering doctorate holders employed in US universities and other academic institutions by main activity.

- - - teachings ———— research

(Source: US National Science Foundation, *Science and Engineering Indicators*, 2006.)

in recent years. In addition, although *Science* does occasionally mention names in their breakthrough commentaries, phrases such as "researchers have discovered," "several groups have observed," and "several reports this year" frequently recur. Discovery has become impersonal. There are few defining names or living legends that might resonate in the imaginations of future commentators or inspire the young. We seem to have turned research into a faceless industry. Is it any wonder that young people are turning away from science?

Although **Tables 2–8** describe some exciting new developments, one might question their durability. Many seem ephemeral and much like progress reports. They give the impression that, exciting though they may be, they will be superseded by the next round of developments before long. It is also important to note that the research community that yielded the harvests outlined in **Tables 2–8** was very much larger than that of the first 70 years or so of the twentieth century. As a rough guide to the global situation, the number of postdoctoral researchers at US universities has increased steadily since 1977, and now stands at some *four times* the 1973 number (see **Figure 3**). (This may be a major part of the problem—there are too many researchers. It is possible, therefore, that governments and funding agencies believe that such a huge army can be kept gainfully employed only if they are managed by objectives.)

Table 9: Nobel Prizes, 1997–2006

Field	Number of Winners	Average Age at Award
Physics	28	69
Chemistry	24	65
Physiology and Medicine	22	63
All	74 (avg: 2.5 per Prize)	66

Table 10: Nobel Prizes, 1961–1970

Field	Number of Winners	Average Age at Award
Physics	18	53
Chemistry	15	57
Physiology and Medicine	26	57
All	59 (avg: 2.0 per Prize)	56

The discoveries listed in **Table 1** eventually opened up vast terrains of exploitable opportunities and transformed economic growth. In many respects, they defined the twentieth century. One can imagine that scientists will still be talking about them 50 or 100 years from now, and some could well be *permanent* monuments to humanity's progression. Indeed, they would seem to define the type of research that a TR initiative should foster.

The claim that individuality in scientific enterprise is losing ground to teamwork and consensus is as impossible to prove conclusively as are Adam Smith's attributions to his invisible hand. Proof of such claims is logically impossible, of course, and one must hope that the authorities are wise enough to see the point. Rather than parade endless anecdotes, I will therefore give only one more example as it also illustrates another aspect of the recent changes.

Table 9 gives the number of scientists sharing the Nobel Prize since 1997 together with the winners' average ages at the time of the award. **Table 10** gives the equivalent data for the decade immediately preceding the 1970s. Up to about 1970, the Nobel Prize was commonly awarded outright to a single winner, indicating a general acknowledgment of that person's overwhelmingly important contribution. Disagreements with the Nobel committees' selections were rare. Nowadays, however, the prizes almost invariably go to the maximum number of scientists permitted by the Nobel rules, namely three, and the fairness of a Nobel committee's selections are often debated, particularly if an important contributor has

Average Age at First Award

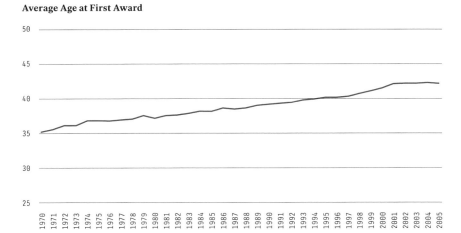

Figure 4: Age at which young investigators win their first independent awards. (Source: US National Institutes of Health, 2007.)

been overlooked. This should not be surprising. When large numbers of scientists pursue similar objectives, it is inevitable that opinions will vary on who did most to get there first.

However, increasing competition often means that younger scientists lose out. Youth has much to commend it, but *potential* is perhaps its most important attribute. As with other abstract qualities, we should expect that consensus measurements will be poor indicators. Not surprisingly, therefore, the age at which young researchers can acquire independent support has increased considerably. Furthermore, as **Tables 9 and 10** show, the ages of Nobel Prize winners have also increased. Elias A. Zerhouni, director of the US National Institutes of Health, spoke in 2006 to express his concern on these matters:

> When I met Marshall Nirenberg, who won the Nobel Prize for Physiology or Medicine in 1968 for discovering the key to deciphering the genetic code and is a researcher at NIH, I went over his history. He started his independent research when he was 27 or 28 years old. At the time, he was unknown. Nobody thought that he had a prayer to crack the code, but he did it four or five years later. He got the Nobel when he was in his mid-to-late 30s. (Morrissey 2006)

This is not an isolated example. **Figure 4** gives the ages at which the National Institutes of Health select new investigators for independent support. **Figure 5** gives National Science Foundation data on the proportion

Percentage

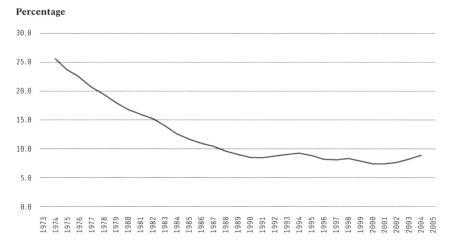

Figure 5: Data showing the decreasing proportion of young (<35-year-old) science and engineering doctorate holders in full-time faculty positions at US universities. (Source: US National Science Foundation, Science and Engineering Indicator, 2006.)

of young (less than 35 years of age) doctorate holders among faculty at US universities. The figures, widely duplicated in other countries, indicate the truly disgraceful treatment of a sector of the community that traditionally has been the source of many scientific points of departure, often in the face of fierce opposition from their seniors. Furthermore, young people today are virtually powerless to do anything about it. The figures indicate that the funding agencies are increasingly overlooking the scientists who are in the most creative periods of their lives. Their creativity may not be lost if their leadership is deferred, but their time spent leading is being severely curtailed.

My choice of the Nobel Prize as a benchmark is appropriate for another reason. For the last five years or so of Venture Research we introduced the informal selection criterion that if we could not imagine an applicant's proposed research eventually winning a Nobel Prize (or one of equivalent rank in the fields that Nobel did not cover),[4] then we should probably not support it. We hoped that this would give further confirmation of the exceptional standards we were looking for and encourage

---

4   Some endorsement of Venture Research's spiritual association with the Nobel Prize ethos is given by John Hurley in his *Organisation and Scientific Discovery* (1997). Hurley reports on a survey of 16 Nobelists' assessments of the most important factors in research. The four regarded as "extremely important" were (1) freedom of thought, (2) choice of work, (3) independence of thought, and (4) freedom to select problems. Two statements with which they "strongly disagreed" were that the frameworks of existing science are incontrovertible and that they work best under the pressure of evaluation. A full acceptance of all these sentiments was at the heart of Venture Research.

applicants to be adventurous. I would strongly recommend it to operators of TR initiatives as hopefully it might lead to them being seen as "schools for Nobelists." Indeed, such schools might appeal to private sponsors as they could have their name associated with them, and could be developed as a possible alternative to TR (see **Poster 9**).

The United States is the largest supporter of research in the world.[5] **Figure 3** shows the dramatic increases that have taken place in the numbers of US doctoral scientists and engineers engaged in research. In contrast, the numbers engaged in teaching have hardly changed, having risen at an average rate of only about 1% per year. The US total expenditure on basic research in all sectors has risen substantially (see **Figure 6**). One might argue that these huge increases have come from the US public and private sectors' supreme confidence in the potential of scientific enterprise, and I applaud their optimism. But similar increases have taken place elsewhere (see **Poster 7**), and in view of the apparently escalating costs of discovery one might wonder, however, whether the authorities believe they are really getting value for money. Researchers are, of course, merely doing what society's proxies are asking them to do—to provide a virtually endless supply of short-term miracles that might justify their support. The inevitable consequence is the rising costs and short life of discoveries in each field as its knowledge base becomes ever more refined. Furthermore, the longer the authorities continue with these policies, the more budgetary demands will increase, and the greater will become the need to fine-tune selection procedures to find the "best" of them. Thus, we are becoming trapped in an ever-tightening vicious spiral.

Some commentators claim that all the easy discoveries have already been made, so we should expect, therefore, that new discoveries will become increasingly harder and more expensive. However, it is astonishing that they use such a word as "easy" in this context. Hindsight indeed can often devalue. The conception of most, if not all, of the major discoveries listed in **Table 1**, for example, seems to have been agonized and protracted. *There probably never has been an easy discovery.* Nevertheless, the dramatic rise of the US expenditure on R&D (see **Figures 6 and 7**) and similar increases elsewhere would seem to indicate that the commentators might have a point.

5   The United States has 17 of the world's top 20 universities, and employs 70% of the world's Nobel Prize winners ("Higher Education," *The Economist*, March 11, 2006).

Poster 9
—

The A. Patron Prize for Science

The Nobel Prize is awarded "to those who, during the preceding year, shall have conferred the greatest benefit on mankind." However, the Nobel Prizes are awarded for achievement in only a small number of specified fields, arcane and secret selection procedures are used, the number of recipients is arbitrarily controlled, and the prizes tend to be very conservative.

The A. Patron Prize would be unique in that suitably qualified scientists would, by their very application for research funding, be declaring their intention of working toward it. No fields would be excluded, and achievements eventually recognized would be of standards similar to those of the Nobel Prizes. Researchers would be selected using the procedures used for Venture Research. Thus, the research would be of the highest international standard, would probably be outside current mainstreams, and would therefore be complementary to any conventional research programs with which A. Patron might be associated. The scientists involved would, in effect, be scholars of an invisible college of potential Nobelists, or whatever name we might give to this new nobelocracy.

The prize itself would be awarded to successful graduates, so to speak, from this college on the advice of a panel of distinguished and internationally respected scientists appointed by A. Patron. The panel would be responsible for determining whether the scientists concerned had indeed transformed *scientific* understanding. These scholars would be a powerful and inspirational force for change. The monetary value of the prize should be exceptionally high—say, $1–2 million for each recipient. The A. Patron Prize would therefore be at the pinnacle of a new and coherent approach to scientific exploration.

The A. Patron Prize would be in a class of its own. It would enhance A. Patron's reputation for farsightedness; would emphasize the value of transparency, fairness, and mutual trust in research; and would help mitigate the intensely bureaucratic procedures that are now threatening to strangle creative talent worldwide. Major advances in unexplored fields can provide unpredicted solutions to current problems and to those of which we are unaware. The prize would thereby add to A. Patron's reputation as a person (or a company) that fully understands the subtleties of human progress.

Expenditures, Millions of Dollars

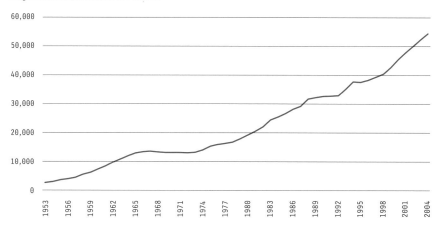

Figure 6: Total US real-terms expenditure on basic research in all sectors and from all sources, including federal, industry, and universities, in constant 2000 dollars. For comparison, the total US expenditure on all types of R&D in 2004 was $288,419 million (2000 dollars). (Source: US National Science Foundation, *Science and Engineering Indicators*, 2006.)

Millions of Dollars

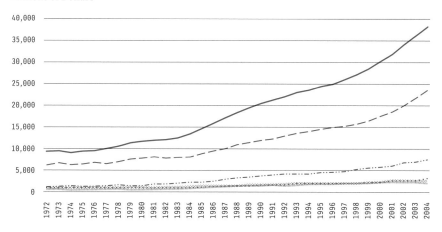

Figure 7: US real-terms expenditure on academic research by sector(s) (2000 dollars):

—— all sources	⋯⋯ industry		
— — federal government	academic institutions		
—··— state/local government	other sources		

(Source: US National Science Foundation, *Science and Engineering Indicators*, 2006.)

I do not agree. As I see it, the increases are evidence of diminishing returns. I am reminded of the joke about the schoolboy who claimed that spiders' ears were in their legs. To prove it, he produced a spider, yelled at it, and watched it run away. He then caught the spider and cruelly pulled off its legs. He yelled at it again, but this time it stayed put. "You see, it can't hear me now," he said. I do not mean any disrespect to commentators or schoolboys, but their logic seems rather similar. Recall the remarks made to the young Max Planck by his supervisor around 1880 that physics was essentially a completed science so he should find some other field to work in. Such remarks are understandable, of course. Coming so soon after the major advances made by such great physicists as Boltzmann, Faraday, and Maxwell, Planck's adviser might have thought there was nothing much left to do.

However, then, as now, major problems abound for those who will look for them. On the atomic scale, we do not understand gravity and its relationship with Nature's other forces. At the other end of the scale, current dogma maintains that the universe was created at some specific moment some 13.6 billion years ago. It then requires an almost immediate but transient expansion at speeds vastly exceeding the velocity of light in order that predictions from current theories can be made compatible with what we actually observe.[6] Furthermore, cosmological thinking implies that we are able to "see" only some 4% of the matter in the universe. The other 96% is hidden from our direct view, which means, of course, that we know precisely nothing about it. One might conclude, therefore, that either we do indeed know very little about the universe we inhabit, or that current thinking is wrong or seriously flawed and we simply do not understand what we observe. In these hugely uncertain circumstances, one might expect that every conceivable avenue was being explored. However, the pressures of consensus are placing severe constraints on the approaches that will be funded. They not only come from bureaucracy: scientists themselves can also contribute. This is inevitable, of course, when the funding agencies encourage them to identify what seem to be the most promising approaches in their fields. I have already mentioned that the Big Bang is virtually the only cosmological show in town. As a further illustration, Lee Smolin, in his *The Trouble with Physics* (2007), points out that string theory—which is concerned with understanding the very fabric of space and the structure of all matter—now has such a dominant position that "it is practically career suicide for young theoretical physicists not to

---

6   Apparently, the expansion of the universe during the brief inflationary period is required to proceed at a speed exceeding 1,020 times the velocity of light.

join the field," despite the fact that, as Smolin carefully explains, it has many serious defects. Mavericks without independent funding, however, must conform or select another topic from a funding agency's à la carte menu, and not only in physics.

Current cosmological theories might finally be confirmed, of course, in which case any intellectual discomfort would be irrelevant. Nature is the ultimate arbiter. Meanwhile, they lack elegance in my view and have a somewhat Heath Robinson "what do we have to do to make the theory fit the data" appearance. Sadly, however, many people regard such esoteric subjects with indifference perhaps because they think they have little relevance to everyday life. But each one of us is embedded in the universe that cosmology, for example, seeks to describe. The fabric of the universe is the fabric of every one of our cells and neurons, and our very consciousness. We are not disinterested observers.

Turning to more obviously relevant issues, our understanding of many microorganisms is poor. Thousands of bacterial species in the soils and the oceans have never been studied. Indeed, it is slightly worse than the situation in cosmology—some 99% of these organisms are unknown. Thus, we know virtually nothing about the overwhelming majority of species we intimately share our environment with. Despite enormous and rising expenditures on research, viral and bacterial infections are still major killers. Some important bacterial infections can be controlled, but we do not understand the processes that drive mutation and thereby severely undermine that control. For viruses, we do not understand the phenomenon of latency by which, say, an otherwise lethal virus can live peaceably for long periods within a living host before finally wreaking its havoc. Mental health is another major cause of concern. To give only one example: Anders Wimo estimated in 2006 that worldwide costs for dementia care were $248 billion a year.[7]

Medicine is understandably concerned with pathological states, but I doubt that we will see substantial progress toward controlling them, whatever their source, until we understand much more about what constitutes good health.[8] Nature's approach to health care is to tackle any problems at the earliest possible moment. Thus, instabilities that can affect human metabolism at every level from the molecular to the whole body, or attacks from bacteria, toxins, and other agents, are usually nipped in the

---

7 Presented to the 10th International Conference on Alzheimer's Disease and Related Disorders, Madrid, July 2006.

8 Chinese medicine, which is far older than Western medicine, maintains that the processes of a healthy human body are interrelated and constantly interact with the environment. Its practitioners strive to understand these interactions so that they can also treat and prevent illness and disease. Western medicine clearly has much to learn from these coherent approaches.

bud by feedback mechanisms or immune systems before they can exert control. Few of these responses are understood. When they fail or have difficulty coping, we become ill. However, medicine's traditional approach has been to treat symptoms, sometimes using aggressive and indiscriminating techniques that can have unpleasant side effects. Symptoms take time to develop, of course, and might be a sign that Nature's procedures need help. Happily, most of us are healthy most of the time, but how can we expect to treat disease and other anomalies if we do not really understand what being healthy means?

Commentators rarely seem to associate today's dearth of major discoveries with the loss of freedom, but freedom is as essential to scientists as legs are to a spider. We seem to live in a universe of infinite complexity. When giving talks, I often like to ask the rhetorical question: "What proportion of what there is to be understood—say, in a thousand years from now—do we understand today?" My answer is: "Very little." There does not seem to be *any* field we can justifiably claim to fully understand, and in these circumstances, there are likely to be many that we do not even know that we do not understand. Nevertheless, by the virtually exclusive use of peer review in the selection of new grants, a notoriously conservative process that few, if any, members of the Planck Club would have survived, the funding agencies imply that all the major discoveries have been made.

If, indeed, many still await us, should we be confident that current policies will find them? As things stand, that seems unlikely as the odds are stacked against individual genius and flair. As **Table 1** shows, my short list could have omitted the scientists' names as their great discoveries are inextricably linked with them, thereby confirming, if confirmation were needed, the essential role of individuality in research. With the exception of buckyballs[9] it is hard to see the evidence for individuality in **Tables 2–8**. The achievements of the past 16 years certainly represent significant progress, but those responsible are usually nameless and unsung, as if only the progress itself needs to be celebrated.

I have long been amused by statements such as "Hampton Court Palace was built (or rebuilt) by Henry VIII." He did no such thing, of course. Numerous architects, engineers, and artisans toiled to bring into being what they thought he wanted despite the many difficulties. We remember their sponsor but not their essential creativity. Today, the

---

9   In 1985, an experiment to unravel the carbon chemistry of red-giant stars led to the discovery of a new allotrope of carbon—C_{60} or buckminsterfullerene, colloquially called "buckyballs"—by Robert F. Curl Jr., Harry W. Kroto, and Richard E. Smalley. Their work was indeed highly individualistic. The problem they wanted to address was not seen as important, and they had to resort to funding some of it themselves. They were awarded the Nobel Prize for Chemistry in 1996.

authorities who control the purse strings seem to have forgotten that great discoveries come from similar workers *who defied the rules*. They must defy them, of course; "the rules" merely being the fruits of past experience. In science, such workers are unique. They have no competitors. The word "best" should not be applied to them. Unfortunately, current strategies lead to cautious, collectively endorsed, and carefully managed research objectives; nonconforming individuals get short shrift. We need a catalyst that will bring about a reversal of that trend.

A TR initiative can be that catalyst. As I shall discuss in the next chapter, the additional funding required would be tiny, but hearts and minds must also be changed, not to mention customs and practices. However, we can draw encouragement from the fact that the transition from a regulated and controlled environment to one of freedom has been made at least once before. As World War II approached its end, scientists everywhere began to think of the future and the restoration of freedom in research. **Poster 10** contains the full text of a letter to *The Times* written by Sir Henry Dale, president of the Royal Society and Nobel Prize winner in Physiology or Medicine, dated August 7, 1945, the day after the atomic bombing of Hiroshima.

Poster 10
—

Science in War and Peace; The Atomic Bomb; Secrecy or Freedom of Research

The following letter, addressed to the editor of *The Times* (of London), was published in 1945:

> Sir, It is a pity, in one way, that the first news of a great event in science had to come at the disclosure of a war secret and to deal chiefly with the "atomic bomb." The world, however, has so arranged its affairs hitherto that we cannot, in fact, imagine that this tremendous realization of a scientific prediction would have been so soon achieved for any other purpose.
>
> The connexion has also the advantage of making mankind face, promptly and squarely, the issues involved. The release of atomic energy, now an accomplished fact, can either destroy civilization or im-

mensely enrich its possibilities; the choice is clearly before mankind and those who guide its destinies. It is everybody's concern and the statesman's supreme responsibility, but I do not believe that the special right of scientists to be heard in its discussion will be challenged. There has been no opportunity for a national, still less an international, exchange of scientific opinion. The world community of science, for the revival of which we now look eagerly, ought to have its say. Meanwhile, may I offer some observations on my individual responsibility, but with a conviction that some other scientists would agree?

This achievement, of all stages, has been the greatest of war secrets, kept with a magnificent loyalty. The scientists concerned will remain loyal to that duty; guarding closely what still has to be kept secret till the war with Japan is finished. Then, I believe, they will all wish to be done with it for ever. We have tolerated much, and would tolerate anything, to ensure the victory for freedom; but when the victory has been won we shall want the freedom. I believe, further, that the abandonment of any national claim to secrecy about scientific discoveries must be a prerequisite for any kind of international control, such as will obviously be indispensable if we are to use atomic energy to its full value and avoid the final disaster which its misuse might bring. If it be objected that this would be incompatible with military secrecy of any kind, I would be bold enough to ask whether that is not already useless. If armaments are to be used only for the international policing of aggressors, what use have we for national secrecy? And have we not, on the other hand, had sufficient experience of the futility and the danger of pacts and agreements which impose quantitative limits on known and obsolescent types of armament, and leave the right to qualitative improvements and new developments in secret?

In any case, is it not obvious that, with the closing stages of this war, scientific discovery and invention are becoming the essential combatants? Science, an unwilling conscript, is becoming the direct agent of undiscriminating devastation at long range, needing only a minimum of military apparatus or personnel—witness the German V weapons and now the atomic bomb. If we retain military secrecy this will be concerned more and more then, with scientific discovery and ever less with fighting men and their equipment. It is bound, if we tolerate it, thus steadily to widen its encroachment and strengthen its hold on the freedom of science. Soon, under such conditions, many of the scientists of some country at peace would find themselves in secret competition with those of another, as ours were ready to be with those of Germany in war, as to which could earlier elaborate the means of annihilating the others, and their countrymen and their country with them.

There, I suggest, lies the threat of final disaster to civilization in place of the measureless enrichment of life which a free science could offer the world. Nor is the moral factor to be ignored. The true spirit of science working in freedom, seeking the truth only and fearing only falsehood and concealment, offers its lofty and austere contribution to man's moral equipment, which the world cannot afford to lose or diminish.

—H. H. Dale, The Royal Society, August 7, 1945

During the war, the US president (F. D. Roosevelt) had set up the powerful Office of Scientific Research and Development (OSRD) to coordinate scientific research for military purposes and to bring the American scientific community into the war effort. Run by Vannevar Bush, it oversaw projects covering the entire scientific spectrum, including the development of radar and the Manhattan Project for the production of an atomic bomb. Thus, for the first time in US history, substantial federal funds were channeled into the universities.[10] Toward the end of the war, President Roosevelt asked OSRD for advice on how peacetime research might be structured. Bush's influential report to the president was concerned mainly with government support of pure research. Bush said:

Many of the lessons learned in the war-time application of science under Government can be profitably applied in peace. The Government is peculiarly fitted to perform certain functions, such as the coordination and support of broad programs on problems of great national importance. But we must proceed with caution in carrying over the methods which work in wartime to the very different conditions of peace. We must remove the rigid controls which we have had to impose and recover freedom of inquiry and that healthy competitive scientific spirit so necessary for expansion of the frontiers of scientific knowledge.

10 Before 1940, the US federal government had generally limited its support for research and training in academic institutions to agriculture.

Scientific progress on a broad front results from the free play of free intellects, working on subjects of their own choice, in the manner dictated by their curiosity for exploration of the unknown. (Bush 1990, 12)

The most important of Bush's postwar recommendations was a proposal for a new independent agency, originally to be called the National Research Foundation. However, the new president (H. S. Truman) insisted that it should be subject to direct presidential control, and vetoed the bill in 1947 after Bush had helped successfully to negotiate its passage through Congress. But Bush did not give up, and over the subsequent three years lobbied hard to unblock the opposition, no doubt drawing hard on his huge reputation and prestige won during his Herculean efforts at OSRD. The National Science Foundation was finally created in 1950, but as responsibility for medical and long-range defense research was excluded, it was but a shadow of what Bush had originally intended. After the war, the National Institutes of Health[11] became the focus of federally supported medical research; other agencies, notably the Office of Naval Research (ONR), took responsibility for defense. The ONR also supported freely chosen and usually undirected basic research in the universities, and many thought it became a model for the newly created NSF.

In contrast, Britain set up the Advisory Council for Scientific Policy, chaired by Sir Henry Tizard. Not surprisingly, research in the nuclear, electronics, and aviation fields was especially favored in the postwar expansion, but otherwise the prewar institutional arrangements were fully restored. Thankfully, they included the so-called dual-support system for academic research, the long-established arrangements by which specific research support was separated from the funds needed to provide for "well-found" university laboratories. This latter support, supplied by the prosaically named but highly influential University Grants Committee (see Chapter 5), meant that academic staff knew that they could usually count on a modest but unquestioned supply of funds to test their embryonic ideas.

The pathways by which the United States and Britain reached postwar scientific normality were different, therefore. But the common factor was that freedom was generally restored, thanks to the tireless efforts of people like Bush, Dale, and many others who passionately argued that although governments should continue their support of academic research, they should not attempt to direct or control it. Luckily for everyone, their

---

11 In 1948 Congress authorized the National Heart Institute to join the decade-old National Cancer Institute, and the former National Institute of Health became the National Institutes of Health.

advice did not fall on deaf ears. Policies based on noninterference with academic freedom flourished everywhere, but I have concentrated on the United States and the United Kingdom because in the immediate postwar period these governments led the Western world in their support for research enterprise and increased their budgets at close to exponential rates until the 1960s. In part, the expansion was fueled by another unforeseen development, the Cold War, as was pointed out in a report to the US Congress:

> The Cold War had an indelible effect on the scientific enterprise, as it provided a compelling rationale for research funding. Indeed, federal research dollars poured into science and technology during this period. The entire enterprise grew; greater numbers of research universities sprung up, more graduate students were trained to become scientists, and entire industries based on new technologies were founded. (*Unlocking Our Future* 1998, 9)

Indeed, academic research took off everywhere on a wave of postwar euphoria, perhaps also inspired by the can-do attitude war stimulated and encouraged. However, the richness of its success also brought the expansion that led to the funding crisis around 1970. Freedom's erosion followed, not this time from any military imperatives, hot or cold, but from the imposition of such well-intentioned considerations as fairness, priority, and relevance, together with the second-guessing censorship imposed by peer review. The harvest from these top-down interventions has been conformity, cascades of technology-led developments, and extensive duplication of objectives on a global scale. Today, Dale's "true spirit of science" is not only being diminished; it is under serious threat.[12]

Ideally, I would like to see a huge controversy on how the wondrous processes of discovery might be freed from bureaucracy's grip. It would mean that the problems had finally been widely acknowledged, and in any case, diversity is usually a good thing. Luckily for fundraisers, TR's costs should be low because its importance is in a similar category to that of vitamin C in a healthy diet. Only trace quantities of that vitamin are required for survival, but on the other hand, taking none at all would soon terminate a person's existence. We must have the catalytic stimuli that TR provides, therefore, and our institutions should make the necessary minor

---

12 J. M. Ziman, in his 1983 Bernal Lecture, said:
Personal discretion in the choice of research problems is now severely limited, even in the university sector, because most projects are now funded by outside agencies. Tension between the individualistic norms of the academic tradition and managerial traditions derived from the industrial tradition has made research an ambivalent profession.

Proportion From Industry

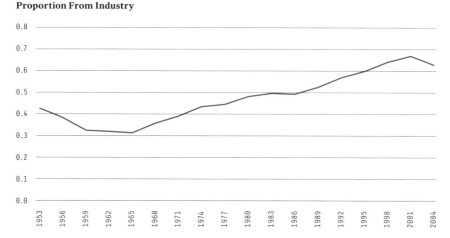

Figure 8: The proportion of US R&D supported by industry. (Source: US National Science Foundation, *Science and Engineering Indicators*, 2006.)

financial adjustments to make it happen. Despite the urgency, it will not be easy. The authorities delayed their acceptance of vitamin C's essential role for several decades after its effectiveness had been scientifically proved. We may not have that time for TR.

As Bush's and Dale's and many other senior scientists' contributions were essential in restoring freedom's role in the postwar years, their successors will probably also need the assistance and support of the nobelocracy, academicians, politicians, industrialists, and other prestigious leaders before our institutions finally become convinced that radical action must be taken. Industry in particular could renew the strong leadership in basic research it had only 10 or 20 years ago. As **Figure 8** shows, US industry has provided a large fraction of the nation's research funding for over 50 years, and it would require only a minuscule shift in present policies to allow the sponsorship of TR either by individual companies as BP did in the 1980s or in consort with others.

Such companies as Bell Labs, IBM, and GE once made legendary scientific contributions.[13] However, US industry support for academic research has declined in the past few years (Britt 2006). The universities' recently acquired penchant for aggressive protection of intellectual property, even when it arises from industry-funded research, may be one cause. Another may be that academic research today is not always distinguishable from

---

13 Listing only the ultimate accolade—the Nobel Prize—Bell Labs have won six Nobel Prizes shared by 11 of its scientists; IBM, three shared by five scientists; and GE, two, each won unshared.

industrial research. The urgent need for publishable and citable results, yet another byproduct of the "law of unintended consequences," produces much the same pressures on academic researchers as the search for saleable products does on their industrial cousins. In these circumstances, it is hardly surprising that industry is losing interest. A TR initiative and the unique perspectives it stimulates might therefore catalyze a revival of industrial interest in academic research to the benefit of both institutions.

Paradoxically, however, their modest costs might mean that the authorities relegate decision-making on TR initiatives to the echelons normally responsible for low-level investments. *That would be disastrous.* Such relegation would bring them into direct competition with investments offering more immediate and certain rewards. It would also mean that those to whom the responsibility was transferred would probably lack the authority to enforce a relaxation of peer review's relentless grip, with the danger that the looming specter of the Planck test would be either fudged or ignored. Despite its low initial costs, decisions on TR must therefore be taken at the highest possible levels—those once occupied by Bush and Dale, for example. The national implications would certainly justify it.

Technology has driven the world since civilized life began. Hunter-gatherers needed guile and courage, but without technology they could not have made the transition to farming, and we would not be civilized. Over the ensuing ages, new technologies ebbed and flowed with societies' toleration of dissent until, as is well known, the Industrial Revolution opened the floodgates. Thanks to Solow, we now understand why living standards in the industrialized nations increased substantially even though populations have been exploding for the past couple of centuries. Not surprisingly, therefore, technology has a special place in the hearts of the authorities. It can enable them to rule or to conquer as they choose.

Technology's primary, if not exclusive, source throughout most of history has been trial and error driven by pioneers dissatisfied with the status quo. Until the nineteenth century, science's role was minuscule, but technology without an understanding of the underlying science is like exploring with one's eyes shut. In these circumstances a technology may or may not work, but one might have no idea why. But science works best when scientists seek to banish ignorance, particularly if they have identified it. Their motivation is then at its highest, whereas struggling to understand how someone else's technology might be improved or extended—"services science" in the 2007 jargon—would rarely be as satisfying.

Support for disinterested science will always be difficult to sell because the full significance of what we do not know is often not easy to recognize. It can take great scientists to do that, as Planck and colleagues have proved. Having done so, however, and derived an apparently viable way forward, one would be faced today with the inevitable question—"Yes, but what good will it do?"—and convincing answers might only rarely be available. Fortunately for everyone, the need for such support could be taken as read until fairly recently. Science's intellectual contributions were fostered in their own right because, to say the least, they enriched our culture. Thus, science's soils were tilled independently of the technological crops that might be grown on them. When the new technologies the Planck Club inspired came along they could flourish, as could economic growth.

The post-1970 academic changes have already been described, but post 1990 similarly constrained thinking began to affect industry. On March 3, 2007 *The Economist* published a well-documented briefing on the shift from research to development, particularly in the big companies. Mammon will have it no other way, it seems. That covetous god has always stalked the marketplace, of course, but would now seem a frequent visitor to the academic sector. Industrialists presiding over R&D's loss of the ampersand, as *The Economist* succinctly puts it, might therefore have a rude awakening when, as they have long been accustomed to doing, they turn to academia for new scientific insights. It may simply not be available in the quality or breadth they look for.

Fresh approaches are now essential. The urgent need is for a few influential leaders to recognize that policies aiming to maximize efficiency will work only when efficiencies can be measured reliably. At the margins of intellectual endeavor where scientists are grappling with difficult or intractable problems the authorities should expect such policies to fail. They might even be the worst possible policies in such circumstances. Until recently, we had visionary leaders, and academic research was wisely left to look after itself. In meritocratic societies vision was once an essential qualification for leadership. But the environment has changed. An industrial leader's prospects are now strongly influenced by the quality of his or her company's quarterly reports. Myopia rules.

Philanthropists also tend to focus on high-performing charities over similar timescales. There are, of course, many who generously donate their resources to alleviating problems in such areas as AIDS, TB, poverty, and education, or to fund deserving institutions. However, we now need a new type of philanthropist—one who fully appreciates the potential value of the intellectual dimension and is prepared to give carte blanche to Nature's ambassadors (see next chapter) who would operate TR initiatives. Indeed, this is in effect the relationship I enjoyed with BP during the 1980s. However, it is a very tall order even though only relatively modest funds are required. Mutual trust is essential, and it is possible that philanthropists might feel more comfortable about taking this radical step if third parties such as, say, national funding agencies were also involved.

4 Searching for Planck's Successors

It is necessary from the outset to deal with the misconception that the advance of scientific knowledge itself can be directed from the centre. This would be to misunderstand the original and sponta- neous nature of science. The advance of scientific knowledge can- not solely be achieved by the arbitrary selection of national goals and by committing resources . . . to them. Because science is origi- nal it is also unpredictable: neither the provenance of a new idea nor its ultimate applications can reliably be foreseen by scientific policy-makers. The tasks of science policy are of another kind: *to maintain the environment necessary for scientific discovery.*

—Council for Scientific Policy's *Report on Science Policy* presented to the UK secretary of state for education and science, May 1966, p. 2 (Emphasis added)

This quotation is taken from the first report by the UK's then newly estab- lished Council for Scientific Policy. Its illustrious membership comprised senior scientists, vice chancellors, research council chairmen, and manag- ing directors (from Mullards and Shell), and was topped off by two Nobel- ists and 17 Fellows of the Royal Society, no less, including its president. Astonishingly, however, the Council's lucid thinking was demoted to the dustbin within only a few years of its appearance. The top-down pseudo- regimentation of academic research ensued, and not only in the UK. The strong influence of the "center" can now be seen everywhere. Funding de- cisions *are usually made* on the grounds of presumed national interests and efficiency.

In addition, some influential politicians seem determined to ensure that government-funded scientists prove they are always earning their keep. Years spent without making a publishable discovery would seem to be years wasted, according to them. Thus, the unpredictable nature of ex- ploratory science is being disregarded. Planck himself "wasted" some 20 years in fruitless inactivity, and many others were similarly "unproduc- tive" for extended periods—see **Poster 1**, for example.

One of the characteristics of our civilization is that its institutions occasionally descend into collective madness giving rise to such lunacies as financial crashes and recessions. Collective perceptions of value can also occasionally lose their usual constraints. The resultant imbalances can be maintained for a time—civilization's uncertainty principle—but wishful thinking is no match for nitty-gritty reality in the long term. Valuations are usually forced down to earth eventually, the only question being whether the landing will be hard or soft. Soaring crude-oil prices and property values are among current (2007) areas of madness, but perhaps the most important of them all is the madness of managing academic research as if its main purpose were to supply commodities. Had the authorities been so afflicted before 1970 or so, the towering contributions stemming from the Planck Club's explorations would have been excluded as they drew their inspiration from such abstract sources as a hunger of the soul. Initially, neither commodities nor practicable applications were usually in sight. In view of the Planck Club's vital importance, therefore, that madness would seem to be the single biggest threat to the stability of the global economy.

Unfortunately, doomsayers are commonplace nowadays. Global warming, epidemics, famine, pollution, and terrorism are serious threats demanding measured responses, but the media (and others) so often overplay them that we have the new threat of threat fatigue. Real signals could be lost in the noise. However, if despite the signs to the contrary you are still confident, as many funding agencies apparently are, that such independently minded scientists as Planck et al. would in today's world be given sufficient leeway to produce the rich harvest summarized in **Table 1** (and that is only a sample), you would have only the doomsayers' threats to worry about. *But you had better be right.* If not, you would, in effect, be forcing the authorities to cope without the wild cards that can unexpectedly change the rules, without that extra shake of the kaleidoscope that can radically change perspectives, without the economic flexibility major discoveries create: these and many other advantages—real and intangible— are precisely what members of the Planck Club among others provide. And given the right environment, they will do all this without being asked!

Fortunately, some institutions are beginning to see the danger signs. Over the last decade or so, some funding agencies have begun to recognize that the current dearth of major discoveries might be due to their overly cautious policies. They have therefore announced schemes to encourage "inherently high-risk research initiatives." In 2002, the Engineering and Physical Sciences Research Council (EPSRC), the largest of the UK research councils, announced its Adventure Fund. It defined adventurous research as "highly speculative research that challenges current conventions, explores new boundaries or adapts novel techniques to an entirely different field with a resultant step change." Although the EPSRC has ended this initiative, it says that it still encourages adventurous research, but it now deals with such proposals through their usual mechanisms.

However, the funding agencies' caution does not stem from a positive decision to be risk-averse. Rather, it is an inevitable consequence of using consensus (peer review) to select the research they will fund. In 2007, the newly established European Research Council (ERC) made the imaginative public announcement that it would give five-year grants totaling €289.5 million (about $300 million) to the support of most promising young researchers—whom they define as people who have held a PhD "not longer than nine and not less than two years." Any subject might qualify subject to the simple guiding principle that it must be at the "frontiers" of knowledge.[1] The ERC says it is looking for "excellence." However, the announcement continues, "proposals will be evaluated by high level peer review evaluation panels," which will probably ensure that the research chosen will not be radical. In any event, the young people targeted will almost certainly not have tenure. It should be natural to expect, therefore, that they would use any ERC support to enhance their chances of getting a secure job in the future, which is hardly compatible with being adventurous.

The US National Institutes of Health (NIH) Director's Pioneer Award Program is described as being a "high-risk research initiative of research teams of the future." The US National Science Foundation (NSF) has a Small Grants for Exploratory Research program, inviting proposals for "small-scale, exploratory, high-risk research in the fields of science, engineering and education normally supported by NSF . . . (proposals) may be submitted to individual programs."

I applaud all new initiatives as they increase diversity. However, I would suggest that agencies are ill-advised when they encourage researchers to develop "high-risk" or "speculative" research programs. A scientist's reputation is slowly earned and quickly damaged, and the length of a productive career can be all too short. A simple question arises, therefore: Why should academic researchers embark on major quests *if their funding agency expects most of them to fail?* That is what sponsoring high-risk research implies.

Perceptions of risk are relative, of course. One person's risk might be another's challenge. Some people might see skydiving as a dangerous pastime because they believe that the probability of a nasty outcome is high.

---

1   The ERC has published the inevitable *Guide to Applicants*, which runs to 53 pages. Specific page and font size, line spacing, and margin widths *are mandatory*. More importantly, the maximum overheads that the ERC will support must not exceed 20% of direct costs. However, many universities are making strenuous efforts to set their overhead fees at much higher levels. Is it reasonable to expect that inexperienced researchers will be able to negotiate that colossal obstacle?

Skydivers themselves, however, while acknowledging and even enjoying the risks, do not expect that their next jump will be other than successful. They would not jump otherwise. *Instead, they very carefully manage the risks.* As their necks are on the line, they take every reasonable step to reduce, eliminate, or control every identifiable source of risk. As a result, they are confident that their jump will be exhilarating and enjoyable, and that they will land in one piece more or less where they thought they would. On the other hand, if it were decreed that the risks of skydiving had to be managed by those paying to enjoy the spectacle—that is, the spectators had to take full responsibility for anything that might happen—it is most unlikely that any jumps would take place. *Responsibility shared is responsibility declined.* In these circumstances, consensus opinion would opt for safety first as it usually does, and yet another expression of joie de vivre would vanish.

Attitudes to risk can also vary between institutions. I was told a lovely story by a senior executive from Scicon Ltd., then (1984) a subsidiary of BP. His company produced battle simulations for the UK Ministry of Defence, and once tried the experiment of interchanging fighter pilots and ships' captains—the pilots would command the ships and the captains would fly the planes. The pilots discovered that frigates have a huge advantage over fighters in that they do not crash when they run out of fuel. So, when the pilots operated the frigates, they pursued their quarry, submarines in this case, much more aggressively than naval captains would. Unlike naval captains, the pilots were not too concerned about calling up a refueling tanker. The result was that the pilots covered twice the area normally covered by sea captains, and "killed" twice as many submarines.

In research, the adoption of "best value for money" policies means, effectively, that funding agencies give the highest priorities to the projects deemed to have the lowest risk of failure. Researchers must manage the residual risks, but so long as they do not stray too far from the beaten tracks, they can take comfort from the fact that consensus deems these risks to be acceptable and small. Should it be surprising, therefore, that research has become mainly predictable and sometimes even dull? However, before 1970 or so researchers were responsible for every aspect of their work, including the risks. We cannot set the clock back, of course, but one of the most important tasks of a TR initiative will be to encourage intrepid researchers to be as adventurous as possible while retaining full responsibility for everything they do.

Imagine that we could plot all this on a few graphs. We cannot precisely quantify levels of success, of course, but we can illustrate the situation symbolically. In **Figure 9**, I plot the probability that a research project

Probability of Success: Mainstream Research

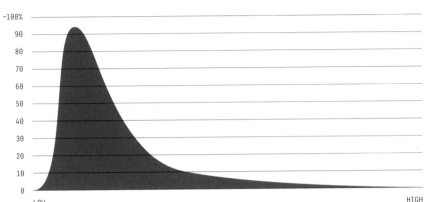

Figure 9: An idealized plot of the possible relationships between a project's chances of success and its potential impact. This plot is for *mainstream* research funded by the conventional agencies.

will be successful against the potential impact that that project might have. "Success" here means that the research meets its objectives. **Figure 9** would seem to accord well with experience. Most projects are expected to succeed but to have a potentially low impact. However, that does not mean that the research was not worth doing. Indeed, mainstream research plays a vital role in consolidating existing knowledge. The long tail indicates that a small proportion of researchers might achieve interesting and exciting results, even though they started out conventionally, as Penzias and Wilson, or Fleming, once did.

In **Figure 10**, I plot the funding agencies' recent excursions into high-risk, high-reward research in a similar way.

These two schematics should cover every type of research the funding agencies currently support. It seems reasonable to ask, therefore, where the work of the Planck Club scientists might have fitted. Their work undoubtedly had high impact, but was it high-risk when they began; that is, was it likely to fail? When Planck et al. were performing their great works, all decisions on risk were the researchers' prerogative. So the question becomes the rhetorical one of how successful they thought they might be, and in particular, whether they thought their work was risky. I cannot conceive that they did. My conceptual abilities are irrelevant, of course. However, my analysis is based on what they *did* and, as far as we know, how they went about it. My thesis is that although they could not know precisely where their "long and labyrinthine paths" would take them, *they were confident that they would prove significant*—hence their enormous

Probability of Success: High-Risk, High-Reward Research

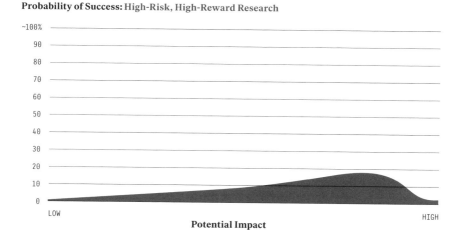

Figure 10: A plot similar to Figure 9, but for *high-risk* research projects as defined by the conventional agencies.

dedication. They had no timetables, except their own, and no one was specifying their goals or urging them on. They would only fail if they gave up. That would be most unlikely, as they clearly believed that if anyone could succeed, they would. As long as they were determined, it would be only a matter of time. I doubt if the idea of risk would have much occupied their minds.

In any event, when funding agencies foster high-risk research, they promote the idea that science is intrinsically risky. It is not. *It is difficult*, but scientists are fully aware of that fact when they begin careers in research. Code breaking, too, can be difficult, but it is not risky; *it is what code breakers do*. They might fail or succeed on a given timescale, but if the code is important, they will continue until they crack it. Scientists are no different. In addition, "risk" is a word usually closely associated with danger. Why on Earth should the funding agencies want to associate danger with difficult problems? If the funding agencies would allow researchers fully to express their individuality some would *choose* to tackle the difficult problems because they would see them as personal challenges.

While no one knows the minds of Planck et al., we modeled Venture Research on what we imagined their inspirations might be. As most of the 26 Venture Research groups went on to make potentially "high-impact" discoveries (see Chapter 7), it would seem that our model is viable, and the high proportion of successes indicates that their research was not high-risk. I illustrate our experience schematically in **Figure 11**.

Probability of Success: Venture Research

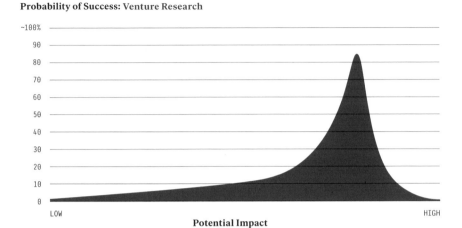

Figure 11: An idealized plot similar to Figures 9 and 10, but based on Venture Research experience.

Venture Research would seem, therefore, to be "low-risk, high-reward," although we did not use such fashionable clichés at the time.[2] We set out from the beginning to re-create the environment of total freedom that members of the Planck Club virtually took for granted. Thus, we *expected* all scientists whom we supported to succeed in their ambitious goals eventually, but we set neither timetables nor goals. If a TR initiative were similarly successful, it would indeed be genuinely transformative.

The way ahead seems clear, therefore. The Venture Research initiative set out to be transformative. As it was very successful the prognoses for TR initiatives would seem excellent *if they are given freedom*. For a national funding agency, the scaling problems mentioned in Chapter 1 would not seem insurmountable because a properly constituted TR initiative should appeal to only a small number of scientists with radical thoughts on their minds. The challenge is to recognize them as there are millions of scientists, and one does not even know which haystack hides the needle. While the major points of departure illustrated in **Table 1** were transformative, few, if any, were extrapolations of what had gone before, nor were they speculative leaps into the dark. Rather, they were the culmination of

---

2   The general reader may not be aware of the funding agencies' routine use of vacuous clichés nowadays. Other widely used examples include "actors," "at the interfaces between disciplines," "benchmarks," "crosscutting," "deliverables," "generic richness," "horizon scanning," "ideas factories," "milestones," "out of the box," "risk appetite," "roadmaps," "sandpits," "stakeholders," and "thematic priorities." None of these vague terms, or the concepts that might be behind them, were generally in use before the circa 1970 watershed. The agencies who use them without apology or deprecation usually indicate a preoccupation with short-term performance and a lack of understanding of what scientific inquiry is about. However, many researchers cannot afford to ignore them as they may contain news of opportunities that may ease their funding difficulties.

long periods of reflective contemplation, or careful experimentation, or both, and a Pasteur-like awareness to be on the lookout for the unexpected.

The achievements listed in **Table 1** may be obvious with hindsight, but a TR initiative must be able to respond adequately to the Planck Club's possible successors *before* they start their campaigns. How can they be reliably identified? Planck was one of the most influential scientists of all time, and few scientists would dream of comparing themselves to him, even though they might go on to make radically new discoveries. However, Einstein in his preface to Planck's book quoted earlier has perceptively and sensitively described a philosophy that in my experience is common to many scientists who have become seriously dissatisfied with the status quo and believe they may have a viable alternative. They, too, might not be overly interested in the power, glory, or other transient rewards major discoveries might bring. Rather, they would seem to yearn for the inner satisfaction that comes from understanding. A TR initiative must therefore be able to respond to these visionary scientists and indeed to encourage them while their great excursions are still at the planning stage.

I do not know of an agency anywhere in the world that as currently constituted could respond to these challenges. Devotion, hunger, and other abstractions are not noticeably high on funding agencies' agendas, but they must be if they are to engage effectively in TR. Some agencies have come to recognize the need for urgent remedial action in attracting adventurous proposals, but they still seem reluctant to provide the freedom to enable that need to be satisfied. Instead, they consult committees of the supposed best scientists and ask for recommendations. But committees can rarely, if ever, be creative. The fact that the best scientists might be involved is irrelevant. It can, of course, provide a defense against criticism as a committee's membership can cloak its conclusions with respectability, but the collective responses of any group, however distinguished they may be as individuals, are usually uninspiring and based on compromise. Scientific advances are none of those things.

So, the tasks of science policy as outlined by the British scientific and technological heavyweights quoted at the beginning of this chapter seem to have moved on. Post 1970s, policy-makers have defined "the environment necessary for scientific discovery" in terms of what the supposed best researchers want to achieve. Funding agencies have then invited researchers to compete for the available funds in each discipline (or combinations of disciplines) according to complex rules usually punctuated with the sort of vague clichés mentioned earlier. Some have recently tried to spice this bland mix with a sprinkling of so-called high-risk initiatives, but the rules have stayed much the same.

It would seem only a matter of time, therefore, before the need for TR initiatives is widely recognized, but how exactly should they be set up? Venture Research experience indicates that in the absence of freedom for everyone—Nature's preferred solution—the funding agency's first step would be to accept that they must restore "the environment necessary for scientific discovery" to something approaching what it was, even if it is only for a small fraction of the community. Thus, the frameworks within which TR initiatives are set up must themselves be transformative. That would be a very big step for many agencies as they seem reluctant to acknowledge that any damage has been done by their present policies, however inadvertent. If they would bite the bureaucratic bullet, however, their reforms must not be formulaic, as any self-respecting prospective Planck Club member would almost certainly disdain them.

Subject to the general acceptance of the definition of TR outlined in Chapter 1, once an agency has appointed the key staff, they should invite them to get on with it. As I have implied, these people would in effect be *Nature's ambassadors*. They would be responsible for deriving the strategies that Nature herself might imaginably be happy with in today's world, that is, for freeing would-be transformative researchers from the ubiquitous rules and regulations, and for encouraging flair. They will succeed only if they can present themselves acceptably to potential Planck Club members. Above all, they must win their trust and respect if these scientists are to share their real intentions with them.

However, and this is a big "however," the diplomatic analogy has weaknesses. In diplomacy, ambassadors may not always be trying to project the same image of their homeland everywhere they serve. Policies in the various ministries (political, commercial, defense, etc.) may have regional variations. There may be hidden agendas. Diplomats should never lie, but they may sometimes be economical with the truth. For these and other complex matters, ambassadors will usually receive their instructions from their governments according to current foreign policies. Diplomats understand these complexities, and this sophisticated service usually works well.

Nature's ambassadors must strive to serve on a much higher plane and should have a positively Planckian dedication to the cause of science per se. Their dedication should also be their source of inspiration as obviously they are no more privy to Nature's secrets than is anyone else. If they are to win researchers' confidence, however, their behavior should reflect what can be imagined we understand about Nature, and one of its most important elements, as Pasteur once famously observed, is that she recognizes only one science.[3] My affectionate creation of the entity I call "Nature" is entirely fictitious, of course, but in reality so are the disciplines, boundaries, or interfaces created by the establishment. Such venerable entities as physics, chemistry, and biology are still nevertheless merely convenient ways of partitioning scientific inquiry. Multidisciplinary research is a vague concept that has meaning only for bureaucrats and apparatchiks—it may not necessarily be a good idea for astronomers to collaborate with zoologists. On the other hand, coherence and flexibility in research are essential; researchers must either have anticipated every contributing factor in their prospective campaigns or be ready to respond to the unexpected as it may arise. Otherwise they will fail.

Nature's ambassadors can be faithful to all this only if the environment within which they operate so empowers them. Thus, the major issues should be whether applicants have coherently and credibly covered all the bases and taken all reasonable steps to ensure that their research would be free from artificial constraints. In a nonlinear world, events have a nasty habit of turning the tables. Indeed, the famous military aphorism "No battle plan survives first contact with the enemy" seems applicable to any sphere, and stresses the importance of considering a range of options. The extent to which these considerations are satisfied will be vital in determining the success of any TR initiative. Indeed, initiatives whose staff is not so empowered would probably be indistinguishable from those we have already.

Funding agencies aiming to create TR initiatives could well have serious difficulties, therefore, as few, if any, fully recognize the unity of science.[4] Fragmentation is the norm.

It is indeed unfortunate that the most significant developments of the newest major discipline—the biological sciences—have taken place during the policy upheavals of the last few decades. Consequently, the agencies have usually emphasized narrowly targeted objectives such as applications and techniques—genome sequencing, for example—because they are easier to assess and prioritize. This is one of the reasons why many breakthroughs announced nowadays look rather like "progress reports," as I mentioned in Chapter 2. The quest for comprehension has suffered as a result.[5] Furthermore, for historical reasons, medical research

---

3 The differing ways in which exploratory research in either the natural sciences or in engineering can lead to surprising and unexpected outcomes were discussed in my *To Be a Scientist* (1994).

4 At a recent meeting, a member of the audience questioned my references to Pasteur's one science with the observation that Pasteur made that remark more than 100 years ago. "We have moved on since then," he said. Perhaps he also regards gravity as a seventeenth-century notion that can now be safely ignored.

5 Sydney Brenner, one of molecular biology's pioneering practitioners, said in an essay published approximately a decade ago: "For many years it was widely held that molecular biology was a completely useless subject, a 'fundamental' science of no interest to those working on practical matters" (1998).

has retained its own hallowed preserve as if our cells and tissues were more important than those found in other forms of life. Such biological sciences as plant, developmental, microbial, and molecular biology are dealt with on an ad hoc basis but separately from the physical sciences. While the current somewhat arbitrary divisions generally work for mainstream research, they would be particularly inappropriate for TR because its main purpose is *to make important and unexpected discoveries.* Unless researchers' minds are fully pasteurized, so to speak, that is, unless they are prepared for every possible eventuality, fortune might not favor them.

In the United States, despite the efforts of Vannevar Bush, the NSF has no responsibility for medical sciences. The NIH has recognized that many important medical advances stemmed from the physical sciences and has consequently opened some of their initiatives to researchers whom they would not normally consider for funding, *provided their objectives are biomedical.* Thus does the specter of Nature's deadly enemy, compromise, raise its ugly head! Many physical scientists who realize that their work could have implications for biology or medicine might make useful contributions under this and similar schemes, but that is not how major discoveries come about. Lasers, nuclear magnetic resonance imaging, and electron microscopy, among many other technologies, have transformed medical practice. They all came from physical scientists, but none had biomedical applications or indeed any other in mind, and so it would be most unlikely therefore that they would have qualified for the NIH schemes had they been in existence at the time. Furthermore, it is not at all clear that their work would have survived scrutiny by the usual NSF mechanisms.[6] The NSF would probably have rejected Townes's proposal for work that led to the laser—had he been required to make one. Even his head of department and fellow specialist did not think Townes's proposal would work (see **Poster 1**).

In the United Kingdom, it is a sad fact that the arrangements for funding academic research are generally more fragmented than in the United States (see **Poster 11**).

---

6   Or by an equivalent organization—electron microscopy was first developed in Germany.

Poster 11

—

The UK's Byzantine Research Councils

The United Kingdom was once the proud possessor of a Science Research Council, the principal organization for supporting academic scientific research. Its virtually exclusive selection criteria were "timeliness and promise." Unfortunately, the move toward greater complexity began in 1981, when it was renamed the Science and Engineering Research Council (SERC), implying thereby that engineering was not deemed a science. In April 1994, the SERC was split into the Engineering and Physical Sciences Research Council (EPSRC) and the Particle Physics and Astronomy Research Council (PPARC). SERC's remote-sensing efforts were transferred to the Natural Environment Research Council, and its biotechnology efforts merged with the former Agriculture and Food Research Council to make the new Biotechnology and Biological Sciences Research Council. In 2007 PPARC was merged with the former Council for the Central Laboratory of the Research Councils, and extracted responsibility for nuclear physics from EPSRC to form the Science and Technology Facilities Council (STFC), based in Swindon. Some scientists have mischievously dubbed the STFC the "Swindon Town Football Club," which would seem nicely to reflect the levels of intellectual input the architects of these confusing changes have hitherto applied.

Medical research is supported from the government purse by the Medical Research Council, and privately at about the same level by the Wellcome Trust. The MRC is the oldest of the research councils—it was created in 1913. In 2007, however, active consideration was being given to a "Single Fund for Health Research," which would seem to involve the MRC being subsumed within the Department of Health. The MRC has been a prolific source of transformative discoveries over the years because, I would suggest, it took determined steps to protect scientific freedom. Subsumed within the Department of Health, it could suffer the same fate as befell the University Grants Committee (see Chapter 6), another one-time staunch advocate of academic freedom.

In contrast, British Petroleum allowed Venture Research to wander freely, but it is a private company that in principle can do almost anything without having to worry too much about treading on academic toes. I have long thought that any new sponsor of Venture Research would have to be private for this reason. Unfortunately, corporate and other private sources of research funds have also increasingly focused their attention on tangible or specific outcomes in recent years. Although several organizations have expressed interest in Venture Research, none have ultimately taken the plunge (see Chapter 5). If the national agencies are not able to allow their TR initiatives to survey the entire scientific spectrum as Nature would surely intend, we could have an impasse.

Optimistically, we can hope that fully free and flexible TR initiatives will emerge soon. In principle, a proliferation of them even in one country should not cause too many problems, although it may lead to confusion among researchers, with such questions as which one they should apply to.

However, although many TR initiatives may be set up, those who create them must not lose sight of the reasons why they are needed. In particular, their staff would still be required to be Nature's ambassadors. If they are to be credible, therefore, it is essential that they be authorized to operate *as if each initiative were unique; as if it were the only one.* TR initiatives confined to specific fields or restricted in any way would be a contradiction and probably a waste of money. They would imply that their sponsors knew the areas where major developments will pop up. But no one can know, not even the researchers.

In large countries like the United States, *it may be preferable* to have several TR initiatives. In my rough estimation (see Chapter 6), such an initiative could eventually be supporting some 108 programs. This number may not seem large by continental standards, but our experience indicates that close relationships between staff and researchers are essential to the building and maintenance of mutual trust and respect. Staff members can play their part only if they are broadly familiar with every project, and among other things, prepared to help with the problems that radical researchers inevitably create. As they are highly individualistic people, one should *expect* that they often will be awkward. However, it is astonishing how intolerant some university administrations can be with researchers whose work does not conform to the norm. In the absence of one's radical view of the world being confirmed, researchers might be out on a limb as far as the rest of a department is concerned, and may not always receive the necessary protection against the routine slings and arrows that bureaucrats love to throw. In Venture Research, scientists regularly asked for our help with such mundane problems as disputes over lab space, appointments, visas, and funding, as well as advice on such strategic issues as research direction and patents. They also seemed to welcome encouragement when things were not going too well.

Sensitive management is people-intensive, and a nationwide US TR initiative would in our experience imply a staff of about 10 scientists with a similar number in administrative support. Although this may not be very different from, say, the staff of an NSF division, a TR initiative would not be administering a program. There should be no specific objectives, priorities (regional or subject), or special initiatives, and little or no peer review (merit review). Indeed, committees or panels should have as little direct involvement as possible. The TR staff will almost certainly wish to seek external advice, but final decisions should be theirs alone. For TR, the intellectual dimension should be paramount. Bureaucracy is, of course, the archenemy of all research, but it can hardly be avoided in the mainstreams if the agencies are to be fair to everyone. For TR initiatives, however, the concept of fairness should take on new meaning. Nature's ambassadors should not strive to be fair to people per se, *but to the very concept of the creative spirit.*

This new approach to fairness also means that every proposal meeting these exceptional standards should be approved. There should be no priorities. This is precisely how we ran Venture Research, but would-be sponsors now seem to view such an apparently open-ended commitment with horror. But it is exceptionally difficult to come up with a proposal that can credibly be argued might qualify the scientist for membership in a future Planck Club or to win a Nobel Prize. Commitments restricted by such severe intellectual constraints would, in effect, limit themselves, as they are so difficult to satisfy, whereas those required merely to support "the best" are effectively unlimited. Even rubbish can be graded. Indeed, it might be difficult to spend even 1% of a large funding agency's budget on TR if Planckian standards are maintained. If expenditure substantially exceeds that level, its strategy would almost certainly be wrong.

It will also be essential that staff members, ambassadors for Nature as they would be, each speak with the same voice, and the smaller their number, the more likely that would be. One would be ideal but hopelessly impracticable. In Venture Research, we opted for the minimum whose integrated experience would roughly span the major disciplines—that is, three or four scientists with a similar number of administrators in support. Proposals in, say, physics, chemistry, or biology should not necessarily come first to their respective specialists, but as intellectual approaches vary among the disciplines we thought we could get closer to Nature, so to speak, by including representatives from each of the major scientific blocs.

We could then learn from each other about how Nature might think. That number was adequate to deal with about 1,000 proposals a year, and a steady-state international community of 25–30 groups of Venture Researchers comprising about 100 scientists. More importantly, it also allowed new staff readily to assimilate the current program, to adjust to the new ways of working, and to learn to think coherently. That might not be possible with either a large team or a large program.

The ambassadors' primary roles would therefore be to encourage and stimulate, and to protect researchers as far as possible from all externally imposed constraints. This means that they should all be generally conversant with the scientific concepts that each applicant brings forward, and clearly the numbers of applications they can manage in these circumstances is strictly limited. They could easily be swamped. To prevent that happening, while keeping the staff of each initiative to a minimum, large funding agencies such as the NSF might eventually choose to have, say, four or five regional TR initiatives with a national director to ensure that each initiative was operating along similar lines. As face-to-face contact is important for TR, such arrangements would also considerably reduce the need for travel.

I acknowledge that these arrangements would be highly unusual for most funding agencies. However, TR initiatives should be exceptional because of the very nature of the problems they seek to alleviate and the potentially revolutionary contributions they could make. Their purpose should not be to serve the profession of science or indeed any social entity.

Our experience implies that the typical cost of a TR initiative (or set of initiatives) would probably be less than 1% of the national spend on research. It may also take some years to reach this level of expenditure because, after decades of mission-oriented research, even potential members of the Planck Club will need time to adjust their thinking to the new opportunities that the initiatives present. Cost alone is therefore unlikely to be an impediment, but national funding agencies may need some remission from the statutory obligations laid on them if TR initiatives are to have the necessary freedom. TR selection criteria should be exclusively scientific—one would expect that Nature might yield her secrets to suitably sensitive, open-minded, and observant members of the human race, but that she would be indifferent to their fields of expertise or such other irrelevant qualifications as age, gender, minority status, or location.

Funding agencies may find it difficult to defend the use of public money on the basis of the abstract criteria outlined here. Thus, we may have a serious problem. Once upon a time, funding agencies confined themselves to the most general issues, but now micromanagement is common.

Nowadays, the realpolitiks imposed by vested interests, pressure groups, costs, and the like severely restrict their room for maneuver. This has led to the emergence of substantial differences between what agencies can do and what ideally they ought to be doing. In addition, the 20–30 years spent in overseeing largely mission-oriented programs may simply have generated more bureaucratic inertia than can quickly be dissipated. These difficulties mean that we may have to look elsewhere than government for TR initiatives, but industrial companies are also having their problems, and philanthropists seem reluctant to provide the unconditional support Planckian researchers need.

It may be attractive, therefore, to explore the possibility of combining the vision and courage often found in the private sector with the commitment to duty found in the public. Public-private TR initiatives funded in partnership could combine the best of both worlds. Flexibility in finance, and efficient access to extensive academic networks and infrastructure could be a powerful combination. One might reasonably believe that Vannevar Bush and Henry Dale would be proud of them!

5   Universities for the Twenty-First Century: The Case for a Fifth Revolution

"If a blunt axe will not sharpen a pencil, ten blunt axes will also fail."

—Edsger W. Dijkstra,
remark to the author, 1983

The university is the oldest of our institutions.[1] Incomparable in terms of the breadth and scope of its intellectual horizons, it is the ideal home for original thinkers on science or life in general. Founded in the eleventh century, it has always been subject to pressures from church or state, but this unique institution has survived because, apart from its functional roles in bringing the next generation abreast of current knowledge and preparing it for the future, it has developed and refined other roles that could hardly be more important or initially less functional. For centuries, society has endowed the university with remarkable degrees of freedom, and this has paid off. Long ago, it was the institution that eventually broke the Church's and feudalism's paralyzing influence in Europe. It pioneered the idea that men (and at that time, sadly, only men) were to be distinguished only by the quality of their thinking, their social status being largely irrelevant. In his superb *History of the English People*, John Richard Green wrote:

> The smallest school (university) was European and not local. . . . A common language, the Latin tongue, superseded . . . the warring tongues of Europe. A common intellectual kinship and rivalry took the place of the petty strifes which parted province from province or realm from realm. What the Church and Empire had both aimed at and both failed in, the knitting of Christian nations together into a vast commonwealth, the Universities for a time actually did. . . . The son of the noble stood on precisely the same footing with the poorest mendicant among Oxford scholars. . . . Knowledge made the "master." . . .

> If the democratic spirit of the Universities threatened feudalism, their spirit of enquiry threatened the Church. (Green 1909, 136–137)

---

[1]   I include within this catch-all term colleges, institutes, and all other centers of higher education.

Unfortunately, like children given something whose value they do not appreciate, the leadership in many countries today seems to believe that there is nothing special about the university. Its performance should therefore be subject to the same indiscriminate processes of optimization and performance assessment other institutions must endure. This chapter discusses the inherent dangers of that misguided and shortsighted policy, and why we must strive to protect this last bastion of intellectual freedom from the tides of homogenization.

In the sciences, Green's "spirit of scepticism, of doubt, of denial" culminated in the seventeenth century with what has been called the first or often *the* Scientific Revolution (Green 1909, 137). The name does not refer to any particular discovery, but rather to the societal, political, and institutional conditions that combine to create environments conducive to major change. The story of this first Revolution (and of the subsequent three) has been told elsewhere.[2] Briefly, such scientists as Kepler, Galileo, Descartes, Newton, and Bacon shaped its development and argued that scientific inquiry should be conducted in *dialog* with Nature. This is in sharp contrast with the received wisdom of the time that scholars should merely contemplate the universe as Aristotle's followers had advocated successfully for many centuries. That does not necessarily mean that these views went unchallenged but rather that any challenges would fall largely on deaf ears.

The Revolution also produced new types of institutions for the advancement, recording, and dissemination of the new knowledge. The *Accademia dei Lincei* (named for the fabled sharpness of the lynx's eyesight), set up in Florence in 1603, was the first. By the 1660s, permanent national academies had been set up in England and in France, and in other countries shortly after. As academies usually founded journals for propagating their members' works and for establishing the priority of their discoveries, these remarkable developments also left a permanent legacy. As they expanded, they institutionalized the continuing progress of intellectual development, thereby laying the exponentially increasing foundations of modern society. Unfortunately, based necessarily on consensus as they are, they have also made it increasingly difficult for heretical views to be heard. As I shall explain, this is now creating serious problems.

Roger Hahn also described a second Scientific Revolution beginning in the early nineteenth century (Hahn 1971). The increase in the number of scientists and people interested in science led to the establishment of the Royal Institution in London, and a wide range of professional societies, the first of which was set up by the geologists. It also led to the creation of the

---

2   For a review, see Cohen's *Revolution in Science* (1985).

British Association for the Advancement of Science and similar organizations in Germany, France, the United States, and elsewhere that catered to the public's apparently insatiable demand for the latest scientific knowledge. Scientists such as Davy and Faraday were superstars in today's language, and their work had a huge popular following. It is a sad fact that although the scope and importance of science in everyday life has increased immeasurably since their time, public interest has nowhere near kept pace. Except for the occasional bursts of curiosity that follow calamities such as earthquakes or epidemics, indifference to Nature's often mysterious ways tends to be the general rule. This is a problem primarily for scientists, of course. Their experiments put them in direct communication with Nature, so to speak, and it could be invaluable if they could galvanize the public with their insights as Davy, Faraday, and many others once did. But it is not easy. Scientists then could afford the investments of time and energy to make their work more accessible because they had much more security in those days. Those taking on these duties today would be penalized as they would have less time to spend on their endless searches for new funding.[3]

The Third Revolution began as the nineteenth century closed and the twentieth opened. The number of universities increased substantially. In addition to their traditional roles in learning and scholarship, they also became centers for scientific research and postgraduate training. Thus, science became established as *a profession* in which a budding scientist could serve a form of apprenticeship and qualify for the right to carry out independent research. Institutes were created with no direct relationships with universities but operating on similar lines. These included the Carnegie Institute in Washington, DC and the *Kaiser-Wilhelm-Gesellschaft* in Germany—later to become the Max Planck Institutes.

Above all, the Third Revolution marked the establishment of the industrial research laboratory and one of the defining revolutions of the twentieth century. Thomas Edison set the scene. While many nineteenth-century scientists such as Faraday with electromagnetism, Pasteur in microbiology and immunology, and Perkin with dyestuffs searched for and achieved industrial applications for their research, Thomas Edison was the master in this respect. He showed that the systematic and rigorous search for and evaluation of practical and profitable uses for science was a profession in its own right. Thus, he made the remarkable discovery that

---

3   For many academic researchers, their continuing employment is also at stake. Up to about half are employed on short-term contracts based on so-called "soft money," and concern about unemployment is never far from their minds.

one could do original and highly creative work developing new products without necessarily uncovering any new science. That discovery turned out to be as important as any discovery in the natural sciences, and led eventually to the phenomenal levels of economic growth that characterized most of the twentieth century.

The potential of the Third Revolution reached its climax during World War II. In the United States and the United Kingdom in particular, academics made vital contributions in virtually every sector of warfare. Discoveries made in such fields as radar, sonar, explosives, detonators, gunnery, weapons design, navigation, logistics, communications, cryptography, medicine, pharmaceuticals, and nutrition were mainly science-based. Engineers were mobilized, too, and helped to achieve and maintain the huge increases in manufacturing production needed to service the voracious demands of total war. The construction in wartime of the Pluto pipeline across the English Channel to supply fuel to the Allies after the Normandy landings illustrates what can be done by inspired engineering and construction genius.[4] Other supreme achievements included the Manhattan Project to design and build an atomic bomb, a project dominated by academics from both sides of the Atlantic and financed almost entirely by the United States; and the predominantly British Ultra program, by which huge numbers of mathematicians, musicians, chess players, electronic and other mainly academic experts combined their skills systematically to break and read the Axis codes (Enigma, etc.) on an almost routine basis. This latter program was so secret that its very existence could not be revealed until the 1970s. I should also include operational research introduced by Robert Watson-Watt (the radar pioneer) in 1939. It was, in his words:

> The application of the basic scientific methods of measurement, classification, comparison and correlation, to the selection of means for attaining, with the least expenditure in effort and time, the maximum operational effect which could be extracted from the available or potentially available resources in personnel and material. (Zuckerman 1966, 17)

Thomas Edison would have approved! It was operational research that led, for example, to the highly effective switch, early in 1944, from the

---

4   Pluto—pipeline under the ocean—consisted of a three-inch-internal-diameter armored pipe laid initially between the Isle of Wight and the Bordeaux peninsula. It was later extended over western Europe. Ultimately 500 miles of pipe were laid, and delivered a million gallons of fuel a day during the final months of the war.

carpet bombing of centers of population to the destruction of the railway network in northwestern Europe.

No wonder, therefore, that Winston Churchill could say in 1943: "When the fetters of wartime are struck off and we turn free hands to the industrial tasks of peace, we may be astonished at the progress and efficiency we shall suddenly see displayed."

These glittering successes set the scene for what Cohen has described as the Fourth Revolution beginning in 1945. Such leaders as Vannevar Bush in the United States and Henry Dale in the United Kingdom (see Chapter 3) catalyzed an unprecedented expansion of government support for academic research. The expansion brought the universities to perhaps their highest levels of performance, popularity, and prestige in the 1950s and 1960s, but the good times did not last. It was not that the universities had failed in any way—quite the contrary—but that their continued expansion at postwar rates had inevitably become impossible to sustain. The post-1970 reforms followed, with their emphasis on narrow measures of perceived efficiency and accountability. In addition, the universities also had a wide range of social obligations heaped on them, probably the most important of which was the requirement that they open their doors to a larger proportion of the population.

Indeed, US and UK universities have come up with the goods in this latter respect, as can be seen in **Tables 11 and 12**. Figures for many other countries are very similar.[5] As Eric Ashby (Lord Ashby) once famously remarked when these increasing enrollments were being proposed in Britain: "More does not mean worse; but it certainly means different."

Now that the slated increases, and perhaps more, have actually taken place, universities everywhere are indeed not the places they once were. Academics have always been proud of their self-appointed roles as commentators on contemporary society's material and intellectual well-being, and of course, they have occasionally made substantial contributions themselves. No one asked them to derive such roles, and in any case, no one could have specified them or even indicated what they might be. Luckily, no one tried. Over the centuries, societies everywhere seemed to have taken the unspoken decision that these reservoirs of extraordinary talent were worth their keep for that very reason alone. One never knew when they might be needed.

One of the Fourth Revolution's most notable events was Robert Solow's Nobel Prize-winning discovery that technical change was the

---

5   D. W. Strangway has recently published details of current and projected student demand on the global scale, in millions, as follows: 6.5, for 1950; 97, for 2000; 119, for 2005; 151, for 2010; 183, for 2015; 222, for 2020; and 263, for 2025 (Strangway 2005, 98).

Table 11: Enrollment in Degree-Granting Postsecondary US Education Institutions

Year	Public Institutions	Private Institutions
1970[a]	6,371,000	2,127,000
2002[b]	12,750,000	3,860,000

a   Source: *Statistical Abstracts of the United States*, 1972.
b   Source: US National Center for Education Statistics, September 2005.

Table 12: Higher-Education Students in UK Universities and Colleges

Year	Number
1969/70[c]	264,000
2003/04[d]	2,247,000

c   Source: Department of Education and Science, 1975.
d   Source: Education Statistics Agency, 2005.

major agent of economic growth. For the first time, therefore, the significance of the Planck Club's discoveries made under the laissez-faire regime of the pre-1970s could be quantified. It was no longer a matter of opinion.

Nevertheless, this superb record has apparently been set aside. Academics are now made constantly aware of their accountability to their various taskmasters and of the need to justify their existence with respect to a host of extraneous responsibilities placed on them.[6] Such shackles are, of course, all too familiar to people in other walks of life, but until recently the universities had generally been exempt. Sadly, however, the bureaucrats, backed up by the politically correct, have now decided that the universities should no longer be excluded from their tender mercies, justifying their decision on the grounds that the academic sector would thereby be more efficient. At first glance, this might seem sensible. ("First glance" is, of course, the bureaucrat's weapon of choice. Closer inspection is rarely encouraged, acknowledged, or allowed.)

---

6   A list of external issues currently being imposed on UK universities was published by F. H. Berkshire of Imperial College, London and includes "efficiency" gains: time-and-motion study, corporate management structures, directed research, European conformity and centralization, examination structures in schools, funding of teaching in higher education, globalization (education as an international commodity), grade inflation at all levels, quality assurance regulation, requirements of employers, research assessment exercises, research/scholarship/teaching balance, school curricula, social engineering and access, and students as "customers" in an international market (Berkshire 2005). As Berkshire puts it: "A common feature of these issues is a fetish for the demonstration of accountability through the creation of performance metrics."

Unfortunately, the major snags are that efficiency measurements require goals to be specified, and their specification is usually based on external considerations. The vast majority of academics rarely have a say.

It might be surprising that academics have generally acquiesced to all this. Ideally, of course, they should have marshaled their vast intellectual reserves to fight for their independence. However, that would not be consistent with the best academic traditions. Academics hitherto were noted for their individuality. Collective action has never been their forte, and in any case, it could easily be represented as special pleading. Thus, the universities, once largely autonomous institutions for the promotion of scholarship and education, are now progressively being brought into line. In effect, *the universities are being made fit for purpose*, but sadly, that purpose is no longer a matter for academics alone to work out. The university must now satisfy society's perceived requirements, which can be as changeable as governments, or as a company's value on the stock exchange. As a result, the universities today are engaged in an apparently never-ending war of attrition as they struggle to cope with the new objectives and the stifling pressures of homogenization.

Impartiality can be a priceless asset, and the university has always stood ready to offer well-informed, well-considered advice and comment on current affairs. Long ago, as Green and others have pointed out, its high-quality thinking helped break the Church's stifling grip. Nowadays, we have urgent need for penetrating insight on terrorism, on global warming and other real or imagined threats to our existence, and on an increasing torrent of so-called health-and-safety-inspired obligations that also combine to curtail freedom. As ever, such insightful contributions are unlikely to be welcomed initially by the "powers that be," but they would seem to be essential precursors and catalysts for action. However, the university's role is much wider than that; it can also emphasize the positive, which is where scientific freedom comes in. There is absolutely no downside to an expanded understanding.

In principle, the university still has much to offer in these respects. But this option may not long be available if its institutions are obliged to recruit only those staff deemed best to meet functional specifications in fashionable fields. In the consultancy field, the old adage was that successful consultants merely delivered the advice their customers wanted to hear. Do we really want that of our universities? If we oblige them to provide precisely what our various proxies think that we need, according to timetable and balance sheet and at least as economically as the competition, they may have little energy left to do the other things, and the university as we have known it will have ceased to exist.

All revolutions come to an end, of course, and the signs are that the Fourth Revolution, too, is petering out; its energy sapped by the bureaucratic paraphernalia of foresight,[7] horizon scanning, relevance, roadmaps, and the like, none of which were even conceived of when the revolution began. It used to be said that researchers strived to learn more and more about less and less, but that will always be the case without the occasional explosive intervention of scientists who stand existing thinking on its head or enable industry to engage in entirely new types of activity. To make matters worse, many researchers now spend more time struggling to cope with bureaucracy than they do on research. The infrastructure's role dominates.

The Fourth Revolution started out with a blaze of hope, its policies designed to free scientists from the necessary clamps imposed by the exigencies of war. Its success has probably exceeded its architects' wildest dreams, but a fog of unintended consequences now clouds its once panoramic vision.

What should we do? Rising populations and increasing personal expectations mean that society as we know it today could break down if we cannot ensure that global economic growth continues to be positive and sustainable. However, thanks to Robert Solow, we now know that growth is strongly dependent on having a vibrant and creative research enterprise. *That vital link seems to have been forgotten or disregarded.* If its significance were more widely appreciated, a stagnating research enterprise could not then be dismissed as an esoteric problem affecting only pampered academics. It would be everyone's concern.

I suggest, therefore, that the time has come to launch a Fifth Revolution. However, revolutions scientific or otherwise are usually only recognized with hindsight. Indeed, as with major discoveries, they are probably not specifiable. They should arise spontaneously whenever the social or intellectual milieu becomes receptive. That condition does not yet seem fully satisfied. In the past few decades, there have been increasing restrictions on personal freedom, and institutions have become increasingly

---

7   Governments that place high value on foresight initiatives, of which the UK government is possibly the leader, usually base their forecasts on those embryo technologies that seem to offer the best prospects. It is in fact a form of projected hindsight as the discoveries on which these technologies are based have already been made. Advocates of these initiatives should study the amusing exercise carried out by *The Economist* (Feb. 12, 2000, 111), which featured the story of two imaginary investors. In 1900, Felicity Foresight invested $1 in the stocks that *would* perform best in the coming year. In 1901, the compounded sum was invested in the best stocks for the next year, and so on until 2000. She was in many respects a Venture Researcher as she based her investments on potential that she was convinced *would* be realized. By 2000, her investments had yielded the staggering sum of approximately $10^{19}. The other investor, Harry Hindsight, also started in 1900, but invested similarly in the best-performing stocks of *the previous year*, which, of course, were clear for all to see. In 2000, the cumulative value of his stock was $783!

cautious and risk-averse, but these changes seem to be regarded as inevitable consequences of modern life to which we must adjust and do the best we can. But such fatalism could easily lead to disaster. In 2007, the fashionable belief was that the greatest threat to humanity came from global warming. In the 1990s, HIV/AIDS was among the major problems cited. The odds are that by 2017 (or maybe before that), we will also be at the mercy of other terrible prospects.

In a healthy society, such threats ebb and flow as we either deal with them, or hopefully they fade in relative importance. But as many would agree, we should not allow them to become so dominant that actions taken to avert or reduce them seriously affect the quality of life. Indeed, in this respect, humanity's prosperity will always depend on *the creation of new opportunities* rather than the identification of threats. We should not ignore the obvious dangers, of course, and we must be able to respond effectively. However, the twentieth century was characterized by humanity's vitality; its ability to create large numbers of major opportunities, material and intellectual, unexpectedly out of nothing. No committees, targets, or deadlines were involved. That vitality ensured that society had the flexibility needed to prosper despite the terrible hazards, including the unforeseen.

What might a Fifth Revolution comprise? If we could project ourselves forward a few decades, I would hope that people then might be saying its seeds were sown around 2007 when it finally became obvious that:

- Real per capita global economic growth is declining.
- Technology is derived from major scientific discoveries made decades ago.
- Managed creativity can at best produce only what its managers specify.
- Efficiency and accountability can sometimes be the worst possible policies.
- We must begin to create a twenty-first-century Planck Club.

Managed creativity (by which I mean creativity directed toward a specific objective or confined in any way) may sound like a good idea, but as my wizard tried to warn us (see **Poster 3**), we do not understand creativity. In these circumstances, the best we can do is to give those rare individuals who seem capable of transformative thinking the freedom to bring their ideas to fruition. Efficiency and accountability have their places; there is little harm in insisting that researchers who know where they are going and when they expect to get there should justify details of their trips. But putative members of the Planck Club should hold themselves accountable only to Nature herself. It took 20 years for Planck to reach his goal, but he had no idea how long it would actually take nor could he have precisely specified it when he set out. What was his efficiency? The question is irrelevant, of course.

The first stirrings of a Fifth Revolution might be heralded by a serious attempt to set up a transformative research initiative.[8] However, it may not be possible to bring about such a momentous event in one throw. It may need several iterations before a full-fledged TR initiative appears in all its glory because some of the bad habits developed since 1970 or so have become deeply engrained. That most pernicious instrument of consensus, peer review of proposals, will probably be the most obdurate because it has acquired the status of *the* gold standard of research quality even though it fails the Planck test. Unfortunately, the gold is the fool's variety, and a TR initiative based on this flawed standard will also fail. We may need patience, therefore. Although the mandatory use of peer preview has been in force for only a few decades, it seems to have infected the very soul of research enterprise. It may need a few rounds of pseudo-TR initiatives before the funding agencies finally bite this bullet. Let us hope that meanwhile some of us have not slid into the Damocles Zone and possible collapse.

Another clear sign would be a proliferation of attempts to protect the university and its most valuable traditions. Before 1970 or so, its elite status was universally acknowledged. However, more recent policies adopted by many advanced nations, such as the OECD member states, would extend the privileges of a university education to some 40% or 50% of the student cohort.[9] **Figure 12** gives the total student enrollment in the United States since 1967.

British polytechnic colleges once offered a wide range of excellent vocational training. In 1992, however, the UK government abolished the so-called binary system, and reclassified all the former polytechnics as universities, thus blurring the distinction between training and teaching. There should be no question, of course, that anyone who can benefit from a university education should be entitled to one. But our leaders do not seem to have taken into account the effects that massive rises in student numbers will have on the concept of the university as an institution. How do they expect that the university's once automatic association with

8   As this book was being finalized, the US National Science Board published a memorandum, *Enhancing Support of Transformative Research at the National Science Foundation*, which invited the NSF to report on its preliminary plans for TR by August 2007.

9   On average across OECD countries, half of today's young adults now enter universities or other institutions offering similar qualifications at some stage during their life (OECD 2004).

Enrollment, Millions

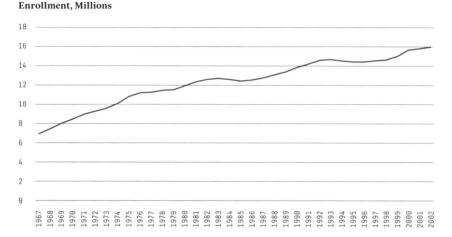

Figure 12: Total enrollment in all US institutions of higher education. The enrollment includes undergraduate and postgraduate students, but does not give their age distributions. In 2001, some 11.4 million were undergraduates while some 4.3 million were involved in postgraduate research. For comparison, the US population of 20–24-year-olds in 2001 was 19.8 million. (Source: US National Science Foundation, *Science and Engineering Indicators*, 2006.)

excellence can continue unquestioned when 50% is the proportion usually associated with average?

It is not at all clear, however, that the new policies offer students a university education per se; that is, the disinterested pursuit of knowledge and understanding. Samuel Eliot Morison, writing in *Three Centuries of Harvard, 1636–1936*, said:

> As long as our teachers regard their work as simply giving so many courses for undergraduates, we shall never have first-class teaching here. If they have to teach graduate students as well as undergraduates, they will regard their subjects as infinite, and keep up that constant investigation (or research) which is necessary for first-class teaching. (Morison 1969, 335–336)

Under the new regimes, it would seem that Morison's "courses for undergraduates" are precisely what will increasingly be on offer, and such ephemeral subjects as "media studies" will proliferate.

In 1988 the heads of over 500 universities worldwide signed the European Magna Charta, marking the 900th anniversary of the founding of the University of Bologna, the oldest in the world. That document included a section on "fundamental principles" (see **Poster 12**). For many of its international signatories, these uncompromising statements on the essence of university autonomy so recently asserted must now seem like wishful thinking—indeed, each one is now threatened. Enabling access to higher education on a mass-produced industrial scale may be socially desirable,[10] but many more teachers will be required to teach (or train) an expanding number of students and they cannot all be involved in research worthy of the name. The inseparability of teaching and research—a necessary condition if university teaching is to be "first-class" according to Morison—would therefore seem to have been abandoned. Academic teaching loads have already increased substantially over the past decade or so. In addition to these, the other social burdens (as opposed to intellectual) being heaped on the universities means that academics must spend so much time meeting deadlines and generally fighting the fires that bureaucrats love to start that they have little time for calm contemplation. We should not be surprised, therefore, if the number left with sufficient energy and inspiration to raise their game to excellent and beyond, as required for Planck Club membership, seems to have fallen close to zero in recent times.

In my fairly extensive experience, most academics have regarded their teaching as an adjunct to their research, an important and stimulating involvement with the next generation that should be taken very seriously but that nonetheless is subordinate to their main mission. That balance is changing to the detriment of research. All except three of the Planck Club short list in **Table 1** did their original work at a university or an academic institution as would the great majority of an extended list. Until circa 1970, they seldom would have been required to justify what they did. That unspoken contract has proved supremely valuable. Academics have been able to make these contributions because the university, having invested considerably in their selection, generally backed their judgments without interference, direction, or comment, with only one minor caveat—that they should discharge their usually modest teaching duties. Unfortunately, the recent changes seriously undermine that contract and the university's potential as an agency for change. As things stand, therefore, we face a future at least as uncertain as it has ever been but without the indispensable and unexpected boosts a full-fledged twenty-first-century Planck Club would provide.

---

10   For an excellent survey of the universities and "the knowledge factory," see the October 2, 1997 issue of *The Economist* (David 1997).

Poster 12
—

European Magna Charta, Bologna, 1988—Fundamental Principles

The following statements are excerpts from the European Magna Charta:

> "The university is an autonomous institution at the heart of societies differently organised because of geography and historical heritage; it produces, examines, appraises and hands down culture by research and teaching."

> "To meet the needs of the world around it, its research and teaching must be morally and intellectually independent of all political authority and economic power."

> "Teaching and research in universities must be inseparable if their tuition is not to lag behind changing needs, the demands of society, and advances in scientific knowledge."

> "Freedom in research and teaching is the fundamental principle of university life, and governments and universities, each as far as in them lies, must ensure respect for this fundamental requirement."

Another noticeable effect of a Fifth Revolution might be the instigation of measures to prevent the university from being subsumed into the homogenized mass of institutions subject to routine governmental control. Almost all universities in the European Union, for example, are government-funded, but even in the United States, which has a large number of private universities, they are not entirely free. Government's indirect subsidies for student support and research are substantial, and private universities are not immune from legislation, of course. It is not clear, therefore, who would instigate such measures. The minimum objective should be to ensure that for at least a subset of universities, academic freedom was no less protected than it was before 1970 or so, an objective that is nicely codified in the Bologna Magna Charta. The university has not yet been terminally or irreversibly damaged, but if we do nothing to alleviate the effects of society's new requirements, it might not be long before its store of exceptional talent becomes worn down by the tedious demands of excessive duty.

In today's materialistic world, the first major question to be settled would be how these Bologna Magna Charta universities should be funded. One would hope that if a Fifth Revolution were actually in progress, materialism might be tempered somewhat, and funds for these relatively few universities would be provided on the basis of excellence *as each university would define*. The major questions would then become: How might we realize this new regime; and just as importantly, how might it be sustained?

The problems would seem more complicated than the provision of personal freedom in research as many universities are state-funded bodies, and so governments must necessarily be involved. Nevertheless, it might also be more tractable because not so long ago it was widely *assumed* that universities should be autonomous,[11] and some states took deliberate steps to ensure that their autonomy was protected. Unfortunately, these steps have now become deeply eroded. One can take comfort, however, from the fact that not so long ago they were in fine shape, and if the lessons of history could be learned, they could provide the basis for a viable solution.

I have strived throughout this book to aim for a global perspective on the issues. After all, science itself is a global discipline. Nationalism has no role whatsoever—it would make no sense claiming to be the first Briton to discover the quantization of energy, for example. Academic freedom has a similar quality. However, perhaps through an accident of history, it would seem that the story of the rise and fall of university autonomy in the UK could provide excellent lessons on what should and should not be done in these respects, and might therefore be worth considering in some detail.

As Daalder and Shils point out, early in the twentieth century the UK government "acknowledged and assured the existence of a large area of free action in which the universities did as they thought right, according to the lights that were interior to them" (Daalder and Shils 1982, 439).

In 1919, it created the University Grants Committee,[12] a body that at least until 1963 rigorously and jealously protected that autonomy. Its membership included the most senior academics and government officials, a formidable gathering in its own right. But perhaps most importantly of all, it was an instrument of the Treasury, that most powerful of government departments. The relationship was indeed special because, not being itself a spending department, the Treasury would probably be content if the

---

11 For a review of the universities in Europe and the United States, see the text edited by Daalder and Shils (1982).

12 For a review of the UGC and its nineteenth-century antecedents, see Tom Owen's article "The University Grants Committee" (1980).

UGC kept to its spending limits and generally did not rock the boat. Thus, the UGC could resist all attempts by most politicians and run-of-the-mill officialdom to interfere, or to closely examine academic expenditures or impose such strictures as "earmarked grants." Thus, universities were free to deal with such labeled offerings as they thought fit. Furthermore, in those heady days funds came to them through rolling five-year grants—the quinquennium—and each university was more or less free to spend their allocation "according to their lights."

In 1963, however, the tanks of a harsher regime began to roll when responsibility for the UGC was transferred to the Department of Education and Science, the same body that had responsibility for the research councils. This apparently innocuous change—another powerful bureaucratic weapon—was imposed not because the UGC was failing—quite the contrary; it was generally held in high regard—but out of an ill-considered urge to rationalize and perhaps to impose more control on this imperious body. The number of UK universities was also about to be considerably expanded following Lord Robbins's influential report (Committee on Higher Education 1963), and change was in the air. Progress—that is, the inexorable tide of events—cannot be halted, of course, but the consequences of these and subsequent administrative changes either seem to have been ignored, or, as some might say, needed hindsight to be fully revealed. However, some provisions—personal freedom in research, for example—are not optional if one wishes to secure a stable and prosperous society, and should have been safeguarded from the outset. Thus, autonomy, at least for a subset of universities, should have been protected so that they could nurture and maintain intellectually challenging environments. These points should not have been difficult to argue. The UGC had demonstrated beyond doubt over many decades that the edict is not the only way of bringing about change. Moreover, their more sensitive and mutually accepted techniques had the enormous advantages of protecting *both* academic freedom and university autonomy. See **Poster 13**.

Poster 13
—

University Governance: An Intelligent Approach to Dirigisme

The following statement is taken from the UK government white paper *University Development 1957 to 1962*, Cmnd 2267, para. 627:

> The major question must be whether the measure of control which the State has entrusted to the Committee (the UGC) is exercised too heavily or too lightly. Some people—not all in the universities—feel that the Committee's influence is becoming too great; others—not all outside the universities—feel it is too light. This balance of opinion is perhaps a measure of their success. But at the same time the Committee do not, and would not, hesitate to exert further influence if it seemed necessary. Only on three occasions in the last ten years have we felt it necessary to take a university directly to task on what we considered to be an inadequate regard to academic standards or national interests. On each occasion, our views have been well received and the response has been immediate. In the main, however, we consider that the universities if given a clear statement of any general or particular situation and a clear indication of what is wanted will of their own volition come to the right decision in the light of their particular circumstances. We, like the universities, recognize that external demands will grow, but we are confident that our main task is to interpret such demands as we feel it legitimate to the universities and to leave the universities to adjust themselves to the changing environment. The delegation of responsibility, we believe, is essential if the health of higher education is to be maintained.

In fact, the change in UGC's governance meant that the dual-support system came under threat immediately—"dual support" is hardly worthy of the name if the same software controls both systems. It also meant that zealous civil servants long denied access to university affairs now became increasingly free. The quinquennium was one of the first casualties, and in the early 1970s the universities had to limit their planning horizons to a single year. In 1989, the government replaced the Olympian UGC with the University Funding Council, and in 1993 with the regional Funding

Councils, one each for England, Scotland, and Wales, and a similar body for Northern Ireland. Furthermore, each of these parochial bodies was now an explicit instrument of government. The bombardment continued with Research Assessment Exercises, Foresight, Quality Assessment,[13] and other bureaucratic impositions that thankfully for scientific enterprise have largely been confined to the UK. Thus, university autonomy became severely compromised, and the climate of fear these bombardments induced was such that the universities offered barely a whimper of protest.

Between 1945 and 1979, UK academic researchers won approximately 41 Nobel Prizes in the sciences—perhaps the ultimate in international quality assessments. That is an average of more than one a year, an extraordinarily high rate for such a relatively small country. For most of this period the dual-support system was still largely intact and among other things provided for "well-found" laboratories that created environments in which researchers could respond quickly to new ideas without the need to involve external assessors. Between 1980 and 2006, the equivalent number was 10, or one every 2.6 years. That fall is not so enormous, of course, and apologists have argued that the reduced figure is still reasonable considering Britain's size. However, six of those prizes were awarded to scientists working at such institutes as the Medical Research Council Laboratory for Molecular Biology, which strived to ensure that personal freedom was protected, and one worked in industry. Thus, the changes have resulted in almost a tenfold decrease in the rate at which Nobel Prizes are won by researchers at UK universities, and for these universities, at least, loss of autonomy would seem to be related to the apparent loss in creativity.

Thus, the United Kingdom has moved from a regime where autonomy was rigorously protected to one where lip service only is paid to this ideal, as is now also the case generally elsewhere. The primary agent of that autonomy—the UGC—was eliminated solely on misplaced ideological grounds. Its efficiency was above reasoned reproach.

Therefore, returning to my global perspective, a general solution to the problem of providing for and maintaining a network of autonomous Bologna Magna Charta universities might be for governments to create buffer agencies such as the UGC that would stand between them and these universities. These agencies would operate to set spending limits, would

---

13 To give an indication of how unadventurous these changes have obliged universities to become, in 2005 the Higher Education Funding Council of England allocated "quality-related" (QR) research funding of £1.2 billion (approximately $2.2 billion) as a block grant to English universities; funding the universities could use as they pleased according to their own priorities. However, a report published in 2007 could not identify a single example of radical departures from mainstream research arising from this funding (SQW Ltd. 2007). That conclusion is all the more remarkable as one of the fund's primary purposes was to foster originality.

generally be responsible for university governance, and should be funded by a powerful nonspending department such as the Treasury so as to ensure, as far as possible, that universities were immune from petty or indeed any kind of external interference. Their membership should include senior academics and others sympathetic to the concept of autonomy, and be renowned for their academic experience and independence of mind. Such agencies might themselves be judged by the extent to which the universities in their care were internationally regarded as excellent.

However, there would still be a major snag. Restoring some of the old universities to Magna Charta status or creating new ones to the same standard would probably be strongly opposed because it would inevitably mean that only a small proportion of universities could claim to be first-class. That could be a serious problem, but there might be another way of satisfying society's understandable wish to provide higher education for a substantial proportion of the population. The limits of secondary education have systematically been extended everywhere over the past century or so. We could take this process to its next logical step by regarding *secondary education* as being complete only after, say, three or four years of study at a normal university. In that regime, therefore, a newly defined higher or tertiary education would begin at a full-fledged Bologna Magna Charta university after this extended secondary phase had been completed, a proposal that would not be substantially different from the measures the UK government is now proposing for England, for example. See **Poster 14**.

Thus, the question of which universities were first-class in this context should not arise. They would *all* be in that category according to some definition otherwise the buffer agency would take the appropriate action. The Bologna Magna Charta universities would then in effect be equivalent to the research universities that we had before about 1970 and would be open only to those who could show that they were qualified. These new universities could be attached to existing universities or be established separately. Student admission should be on merit only as it still is generally for postgraduate study today, but researchers at the Bologna Magna Charta universities should be as free as they once universally were. That is, they would all have access to modest levels of funding to use as they pleased. Hopefully therefore, this provision should restock the reservoir from which new Planck Club members might eventually emerge.

Poster 14
—
Elementary and Secondary Education in England

The time line is as follows:

1880   Elementary education between the ages of 5 and 10 years became compulsory and free.

1918   Age range for compulsory and free education extended to 14 years.

1947   Age range for compulsory and free education extended to 15 years.

1972   Age range for compulsory and free education extended to 16 years.

2005   The government published plans for a new national entitlement. The government's Department for Education and Skills announced that it wants all 14–19-year-olds in the country "to have the opportunity to pursue a course of study where they will learn in a style that suits them and in subject areas which motivate them. Whether in school or college or on an Apprenticeship, we want young people to be working towards qualifications which have real standing with employers and the general public, which prepare them to progress into further and higher education, and which place a premium on mastering the basics." The aim is that the new national entitlement will be in place by 2013.

In the absence of such UGC-like agencies, academic leaders and others would at the threshold of a Fifth Revolution find ways of strongly resisting the trend of turning the universities into service-providing businesses. This threat is becoming real, as William Waugh has written:

Colleges and universities tended to choose their leaders from the ranks of the faculty, often with academic qualifications outweighing administrative skills. To some extent, many academic institutions still follow that model. . . . Now, however, there is increasing pressure to recruit executives from the private sector armed with skills in business management but without work experience in academic administration. As a result, academic planning, budgeting, and day-to-day administration is becoming like the management processes developed for the private sector and increasingly reflects values that conflict with the traditional values of university governance. (Waugh 2003, 85)

As the trend continues, students will increasingly be regarded as customers, and academic research as a supplier of commodities. The commercial pressures on academic research are all too familiar, especially as companies are cutting back on their in-house exploratory research. The pressures from customer-students to provide such user-friendly courses as cosmetics, costume design, and the culinary arts are already increasing.[14] In the circumstances where governments are tending to focus on short-term pragmatism, who, in the absence of an organization like the UGC, would argue that a university education should be as demanding as possible? Customers often go for the easiest or cheapest options.

Alternative models of tertiary education should also be tried. The tacit assumption seems to be that present structures and responsibilities are the only viable options. This is hardly surprising, and is an extraordinary tribute to the longevity of the concepts developed at the beginning of the last millennium. Thus, one usually finds autonomous and relatively independent faculties in say the arts, the sciences, engineering, economics, and other disciplines, although some institutions may specialize in one or more of these broad groupings. Traditionally, most students are young and inexperienced, and the expectation has been that their training should be adequate preparation for a career spanning several decades. That is no longer the case. The demand for continuing education is increasing, but the traditional formula by which students seek to acquire new skills from expert teachers remains much the same. It might be time, therefore, for a reexamination.

Our Venture Research experience indicates that there might be another approach. Our search was for scientists with flair and panache, but as we found in our workshops such people can also be inspirational, which is or should be a vital component of the educational processes. Thus, we found that their novel and defendable perspectives stimulated the attendants—industrial scientists and other experts—to see their own problems in a new light, which in turn promoted interactions from which everyone

---

14   These snippets are taken from a depressingly longer list given in Sir John Meurig Thomas's excellent paper on the impact of these pressures on university curricula (Thomas 2001).

benefited. Indeed, the workshops were so successful that we thought their format could form the basis of a new type of university (see **Poster 15**). It departs from tradition by focusing on education as an *enlightening experience* and deliberately blurs the usual distinctions on status by aiming to foster intellectual interactions between people of comparable standing.

Poster 15

—

Outline Proposal for a New Type of University

Venture Researchers were scientists, of course, but our selection methodology should be generally applicable. In literature, music, the arts, and other fields, intellectual divisions are as similarly arbitrary as those found in the sciences, and they also change with time. The ideas of a latter-day Picasso or Stravinsky would probably struggle as much for acceptance as did those of many members of the Planck Club. Since Venture Research was committed to the idea that intellect is indivisible, its modus operandi could be used to identify those who have a unique, credible, but radical view of the human condition in general.

The proposal stems from these observations. The new type of university would be based *on people*, and would not have departments in the traditional sense. Members of "staff" would be selected for their radical and stimulating viewpoints. They would attend for short periods, say, a week initially, and be selected from a specially appointed panel of such people. "Students" would attend because they felt a need for intellectual refreshment. They would be mature and expert in some field. The relationships between staff and students would be based on mutual trust and respect between accomplished equals.

Students would attend for short periods for which they would pay or be sponsored. They would be from industry, commerce, the civil service, or indeed any walk of life. There would be no examinations and no degrees. The new university's main objective would be to help students develop their own critical and creative capabilities, and to show them that life in general can be much richer than present orthodoxies recognize. In effect, participation would help students prepare themselves for the next phases of their careers rather than train them in a particular skill. Eventually, the aim would be that periodic attendance at universities of this type would become as important as graduation from a conventional university at the start of one's career.

In keeping with its open intellectual approach, the new university should eventually be fully international. There would be a minimum of bureaucracy. It would probably be privately sponsored at first, but should soon be self-supporting. Initially, it could be attached to an existing university or group of universities, and operate for only a few weeks a year. Each course might last, say, one week, with the range of lectures/tutorials covering as much intellectual ground as possible. Student numbers for each course should be small, say, 30 or so, with, say, five to 10 staff. Courses should be residential if possible. Each course would have a director for coordination, and so on.

The new university would catalyze a considerable increase in the range of relationships between the academic and other professional worlds—industry, for example—for which there could be many unpredictable benefits.

The rising tide of events that later became known as the Industrial Revolution began about 250 years ago, and was arguably the most momentous in human history. Although its success has been prodigious, the global have-nots today still outnumber the haves, and there are many problems. We now have a new one—this ultimate of revolutions seems to be running out of steam. For over 200 years, we could enjoy its wild and prolific harvest, which like the seasons could virtually be taken for granted. Major scientific discoveries erupted spontaneously but at an acceptable average rate. Economic growth seemed to follow faithfully and somewhat mysteriously, but no sooner had Solow identified its source than we began to lose our unquestioning confidence in that magical relationship. As soon as they thought they understood all this, our proxies decided that progress would proceed more efficiently if research were focused on bringing about the technical changes we wanted when we wanted them. To make matters much worse, presumed industrial shareholder expectations are increasingly used to determine corporate research policies, thereby further inhibiting flair and audacity, and virtually eliminating the chance arrival of spontaneous discoveries.

The cancerous growth of consensus and other bureaucratic scourges could prove disastrous. In the academic sector, freedom is being curtailed because some influential people fear that otherwise freeloaders will exploit it. In industry, too few of its leaders are prepared to lead and to persuade shareholders, for example, that short-term competitiveness can sometimes compromise profits and the prospects for long-term survival. All too often today, however, understanding is sacrificed in favor of tangible objectives. Understanding is, of course, a wholly abstract concept. I do not know of any civilization that collapsed because it had too much, whereas all too many failed because they clamped down on its supply. Our future is precariously balanced. Our burgeoning problems mean that we have a tiger by the tail, and survival may depend on having all the understanding, flexibility, and audacity we can get. We may learn new tricks thereby, and might even learn to tame the tiger. Many civilizations have eventually ended in collapse, but it is not inevitable. Thanks to the elixir that comes from scientific freedom, we can postpone it indefinitely if we appreciate the value of our most precious asset.

New knowledge usually comes from maverick loners. Society's full recognition of that fact, and all its consequences, could be the most permanent legacy of a Fifth Revolution.

6   Venture (or Transformative) Research: How It Works in Practice

> This gift of seeing things where others see nothing is indeed the mark of scientific genius, a faculty guided by vague criteria, just as are the feats of keen eyesight. That is what the members of the world's first scientific society boasted of by calling themselves the *Accademia dei Lincei.*

—Michael Polanyi,
Knowing and Being, University of Chicago Press, 1969, p. 107

In 1980, Dr. Jack Birks, a BP managing director, invited me to come to BP and set up what they hoped would be a radically new research initiative— the Venture Research Unit—charged with creating completely new types of industrial opportunity. He offered complete freedom, which was rather worrying as I would have no one to blame if I failed! In those days, managing directors of large companies were expected to back their judgment. They were responsible to their boards, of course, but they were usually free to take any action they considered to be in the company's best interests, a justification they could postpone until after the event. They had to be mostly right, of course, at least on the important issues, or they would not survive. The delegation of such powers is now rare in big companies. Committees and focus groups have now replaced individuality, and flair seems to have flown out of the window. That is a pity. The interests of industrial giants once extended far beyond what shareholders might initially expect. Until the late 1970s, BP's research planning was based on a 10-year "look ahead," and BP's researchers were allowed the freedom to generate "new basic knowledge which may have some relevance to the group's future business" and which may lead to "completely new business outlets and to unsuspected solutions to existing problems" (British Petroleum Company 1977, 421). The creation of the Venture Research initiative in 1980 was the culmination of that policy. Sadly, BP was to terminate the initiative 10 years later even though BP acknowledged its considerable successes. BP's reasoning was that it did not want to give shareholders a signal that the company might be considering even the minutest departures from an exclusive concentration on core businesses.

There are some signs today of a reversion to older more confident ways among big industrial companies. GE, for example, is reported to be moving away from its famous "six sigma" policy, which obliged GE managers to concentrate on being consummately good at delivering incremental improvements precisely on time. In a recent interview with *The Economist* (Dec. 10, 2005), the GE corporate leadership is quoted as saying: "In future [managers] will be judged not only by all the usual measures such as return on capital that investors typically care about: they will also be held accountable for helping to save the planet." My hope is that TR initiatives, particularly if they can be set up cooperatively with visionary industrialists, will help to encourage this welcome trend.

In 1980, Birks was responsible for almost all the technological aspects of BP's businesses—exploration, production, engineering, research, and development—a huge portfolio. Not surprisingly, he was somewhat autocratic, a trait he skillfully cloaked with a relaxed and good-humored appearance. However, I have never met anyone who could cut through woolly thinking more quickly or mercilessly. He was also rather devious. To keep me in check, he had arranged that I would be responsible to a specially constituted board of some of the most prestigious scientists in Britain and the most senior executives in BP—a formidable phalanx of establishment leaders. Sir James Menter, principal of Queen Mary College London, a non-executive BP Director, and a distinguished physicist and industrial scientist, chaired it.[1] Menter was a man of very few words; none at all was his usual preference, his silence usually meaning to trained interpreters that he was content. That meant that he and I had interesting meetings, as they say, but I knew that if I could persuade Jim to do something, his authority was such that it would probably happen. Birks thought that Jim and I would make an excellent team because he knew I would never give up. Menter could be obdurate, but if I could convince him, which was often as difficult as what I imagined scaling the north face of the Eiger would be, he could be a powerful and generally unstoppable ally.

British Petroleum offered total freedom, so we could aim therefore to create a unique initiative. As major new industry—that based on the laser for example—is unpredictable, I decided that we would concentrate exclusively on the type of research that might lead to its creation. But global research expenditure on research and development is gigantic (see **Poster 7**), while Venture Research's would not even register on the global scale.

---

1   The other external members were Sir Hans Kornberg, professor of biochemistry at Cambridge (later to become master of Christ's College Cambridge), and Sir Rex Richards, warden of Merton College and professor of physical chemistry at Oxford. Ex officio members from BP were Professor John Cadogan, director of research; the head of corporate planning; and BP's chief engineer.

Indeed, at its peak it turned out to be less that 0.5% of BP's research budget. How on Earth, therefore, could we hope to come up with something that was not already being done? We decided that we should differentiate ourselves by striving to support research as in an ideal world it should be supported. We did not presume that our model would necessarily be applicable to mainstream research. That was not our brief. What we needed was a disciplined route to the unexpected, and we decided that it might come from striving to create a Planckian perspective on the sciences; that is, we would submit our procedures to the Planck test from the outset. In these circumstances, it would not matter if subsequently we found that other companies were doing much the same. The Planckian approach fosters individuality, and just as no two fingerprints are the same, it would be most unlikely that even if others had adopted the same strategy, we would be in direct competition with them.

The Planckian approach might lead to uniqueness, but the adoption of such a high-minded course of action is easier said than done. We might have been expected to slide into expedient ways, especially as that is what some of our critics expected us to do. Thus, we might have chosen a few terribly important high-priority areas[2] and teams of the most prestigious scientists to add luster to our program. We could then have proudly joined the crowded high-tech circuit of conferences and meetings in exotic places safe in the knowledge that we would probably not be criticized because that is what everyone was doing.

Ernest Rutherford once said: "When money is short there is no alternative but to think." It seems a sad fact that if one is awarded a large grant, as might be the case in such fields as nanotechnology or the hydrogen economy, so much money is sloshing around that thinking is not necessarily essential. One can be kept respectably busy managing the funds. So, inspired by Rutherford, we took the line that although BP had not yet revealed the limits to its altruism, they would probably not be very high. We would look, therefore, for researchers who needed freedom above all because the costs of providing that rare commodity are almost exclusively intellectual. They would need funding, of course, but we soon found that the requirements of freedom-starved researchers are usually modest. That should not be too surprising, as the total initial support for the 20 members of the Planck Club listed in **Table 1**, who of course did precisely as they pleased, could have been provided at a one-off cost of about $20–$30 million in today's money, which of course would have been one of the best

---

2   In 1980, the four areas that many expected us to support were biotechnology; electronics, information processing, and robotics; new materials; lasers and optoelectronics.

investments in history! One merely has to derive an efficient means of finding them.

Putting ourselves in what we imagined Planck's shoes might be, and spurred on by Rutherford's dictum, we set out to create an initiative of which both might be proud. We started by trying to eliminate every selection rule imposed since about 1970 that appeared to stand in the way of freedom. This was a long and painful process. After a few agonizing years, our main conclusion was that we should concentrate exclusively on trying to identify people who had recognized substantial flaws or gaps in existing knowledge, and who might know what to do about them. Bearing in mind that the researchers we were looking for are rare—membership in the Planck Club should be rather exclusive—we would make it as easy as possible to apply. Written applications could be made on one page or preferably less, and if researchers needed a quicker reply, they could apply by telephone. With such a low application threshold initially, we would not then feel too guilty when we had to tell almost everyone that their research did not seem to qualify, nor would we have wasted too much of their time. The important word here is "seem." We would add that any researcher who did not agree with our apparent rejection could return at any time with a rebuttal. Thus, however long (or short) our dialog might be, we tried to arrange that we always returned the ball to the applicant's court. The onus would always be on them to respond to the line we had taken.

In effect, therefore, we were creating an *environment in which researchers could select themselves* (see **Poster 16**). We, as Nature's ambassadors, would manage that environment and try to ensure that all who entered should have every opportunity to share their scientific vision with us. Applicants mistakenly turned away could return if they wanted to. Our possible error would then have served to test their determination, an essential Planckian quality that was precisely what we were looking for. We did not set such tests mischievously, of course, but this element had the looked-for "fail-safe" quality, and helped to reduce our worries that our lack of expertise might make us miss golden opportunities staring us in the face. Fail-safe selection procedures would mean that if applicants were determined, and there would be no point of supporting them otherwise, we might have at least one more stab at the gold.

However, there are a few big snags. We were looking for scientists with a Planckian turn of mind, but they might currently be neither well connected nor well known. We decided, therefore, that all proposals should be unsolicited. We were not an established funding agency, of course, so we had to spread the news of our interests "on the hoof" as it were, speaking at as many universities as we could persuade to invite us, to the media, and publishing in those *very* few journals that would permit it.[3] We thought about advertising, but always decided against it. Our reasoning was that the Olympians we were looking for might see it as beneath their dignity to respond to an ad.

Poster 16
—
An Environment for Transformative Research

The prospect for success for a transformative research (TR) initiative will depend on the extent to which we can restore the conditions that prevailed until about the 1970s. That is impossible for every scientist, of course. TR is unpredictable, so we need an environment in which there is a good chance that the tiny few capable of transforming scientific enterprise can select themselves. Unfortunately, that environment must be managed, but its promoters must strive to make it appear as if it were unmanaged, as if nothing had changed.

Some properties of that environment would be:
- "Free" funds (i.e., funds available for use as required).
- No boundaries.
- No deadlines.
- No exclusions.
- No milestones.
- No peer review.
- No priorities.
- No specific objectives other than to understand or explore.
- Researchers free to go in any direction at any time.
- Risks to be selected and managed by the researchers.

Above all, the TR environment must foster mutual trust and respect.

---

3   I do not understand the scientific literature's almost total indifference to what Venture Research is trying to do. It is as if we are being vetoed by their peer-review procedures. I gave a specific example of this in *Pioneering Research:* first *Nature* and then *Science*, two of the world's most influential journals, declined to publish in 2001 a short letter (238 words) from 19 senior British and American scientists (including two Nobel Laureates) marking the 100th anniversary of Max Planck's seminal discoveries, and drawing attention to the problem of scientific freedom in general terms. Neither gave a reason.

The academic research environment had all or most of these properties before the 1970s, although they may not have been explicitly stated. However, the promoters of TR should not be unduly prescriptive about what is and what is not TR. They will need to be alert to unusual proposals and to pursue them rigorously, as if they were bench scientists struggling to understand an unusual observation.

Our message was intended to convey respect—respect for every aspect of the sciences, respect for those who seek to deepen and extend understanding independently of tangible benefit, and respect for the courage needed to tackle challenging and perhaps intractable problems. Our immediate objective was to remove the selection process as far as possible from the routine,[4] at least as far as the serious contenders were concerned. In our case, this latter number turned out to be about 100 or so a year, all of whom we arranged to meet. We then sat down near a whiteboard and talked about whatever science interested them. Almost invariably, the first thing we had to do was to discourage them from describing the possible benefits that might flow from their proposed work (applications, technology, etc.). In this, they merely seemed to be following normal practice by telling us what they thought we as a funding agency wanted to hear. The level of possible funding was another of their priorities. Our response was to ask them to assume that we had an infinite amount of money and could offer freedom to match! With these and other extraneous issues so beloved of conventional funding agencies out of the way, we could finally ask them to tell us what they would like to do that they were not doing now.

As our scientific team was never more than three or four (and often less), our coverage of the huge range of current scientific disciplines was, to say the least, severely limited. However, we soon saw that we could turn this apparent drawback into a strong advantage. Normally, the funding agencies aim to assess whether a team's specific skills would be adequate to meet its declared goals, and whether those goals would be worthwhile. The agency then invites a small number of the team's peers, who are also their closest competitors, to subject the proposal to a microscopic examination against these criteria, which verdicts they can deliver anonymously. By focusing our discussions exclusively on scientific concepts, we would avoid getting bogged down in distracting details. Our perspective could only come from a highly elevated vantage point, of course, but then we

---

4   We got about 1,000 applications a year, but about 90% of those seemed to be from researchers seeking additional funds to extend the work they were already doing rather than freedom to embark on new departures. However, we encouraged them to reply if they disagreed.

were looking for researchers who also had panoramic vision. We knew, of course, that real progress requires firm contact with Nature's undergrowth, so to speak, but we could defer any assessment of their abilities to penetrate that until later, *if necessary*. Most funding agencies focus almost entirely on these mundane, if essential, abilities—which is one of the main tasks of peer preview—so here was yet another departure for our prospective Planck Club members to get used to. But our concentration on concepts created a sensitive test by separating those anxious to impress us with their tactical expertise from the strategists who knew that their tactics would have to vary as their research evolved. We were looking for coherence and flexibility. Uncertainty (as opposed to risk) is *an essential* part of scientific research worthy of the name. It therefore makes no sense to dwell on any one set of techniques or tactics at the outset.

The best of our applicants not only had no difficulties with our novel approach but positively rose to the occasion. But they did not do so immediately. They seemed to want to gauge our seriousness, and more importantly, whether they could trust us before they would tell us what they really wanted to do. This latter point is vital. Why, indeed, should they tell us what they were thinking if our first reaction would be to share those thoughts with their closest competitors as conventional peer review requires. There is no magical formula for this any more than there is for cementing personal relationships in general. We merely talked to them as one scientist might talk to another—inviting them to tell us about their plans, how they had arisen, and to put it all into a global context. Our agenda was to understand what were likely to be difficult concepts, and so we leavened our discussions by talking about the weather, or music, or anything that might put them at their ease and increase the efficiency of communication.

We also needed to build our trust in them. If we were to fund their proposal, it would be with the lightest of touches. They would be free to do whatever they liked with the funds we provided, subject only to the usual accounting constraints. We expected them to address the problems we had agreed on, and there would be a contract between their home institute and BP describing their research in general terms and guaranteeing our support for, say, three years. But we designed the contract so that it would give researchers full freedom of action. We could offer nothing less if our arrangements were to pass the Planck test. We could be happy with these relaxed arrangements only if we trusted them. They would only tell us what they wanted to do if they trusted us not only to provide funds but also to back them when things might not be going well. Once we had made up our minds to support them, we became their collaborators. Henceforth, we became part of their team, writing and shepherding their proposal through Menter's board, and being involved with every aspect of their progress thereafter. Their problems became our problems. We also took vicarious pleasure in their discoveries. In my case, that was certainly an important part of the trade-off in giving up my own research career.

However, there was another important step before we could go forward. Once we were convinced about the research, we needed to satisfy ourselves that the researchers were capable of doing it. To start, we might consult a suitable expert. This is not the same as peer review. We had effectively made up our minds, but wanted to be reasonably sure that we had not missed anything. Thus, for example, we asked for *impressions* on whether applicants seemed to know what they were talking about, and whether the data they proposed to take would give them the information they were looking for. This consultation would often throw up useful suggestions, but in our experience never revealed real problems. One of us would then visit the researchers in their home environment "to kick the tires." They would show us what they were doing and something of what was going on elsewhere in their department. Almost invariably, the fields would be new to us, but we found that it is remarkably easy to judge levels of competence from the fluency with which presentations are made, or questions answered (or avoided). As we would often tour their labs, it was interesting to note other scientists' responses to what had been proposed, and the degrees of respect (or otherwise!) in which the applicants were held. These visits would be low-key, and take a few hours. In our experience, they never raised problems not already discussed and were usually a formality. Nevertheless, they formed an essential part of our "due diligence," and helped give us confidence that we were doing the right thing. The wool might be being pulled over our inexpert eyes, but every bench scientist might have similar worries with unusual sets of data. There is no alternative to constant checking and rechecking, but eventually, one has to back one's judgment. We now have the clarity of more than 10 years' hindsight from which we *know* that our decisions, often taken in the teeth of peer-review assessment, were the right ones. It is indeed ironic that the procedures we regarded as a formality are usually the prime focus of the conventional funding agencies' attention, much to researchers' frustration, especially when success rates are low.

Chapter 7 describes *all* our Venture Research projects running at the closure in 1990. It seems reasonable to conclude that if our researchers did succeed in radically changing the ways in which we think in important fields, and our novel procedures did provide a reliable route to the unpredictable, we might justifiably claim to be acting as proxy for Nature. We could not know that at the time, of course. We thought that they could hardly fail because their lines of attack were unique and flexible, the areas were important but poorly understood, and their dedication was total. They also knew we would back them through thick and thin. We could not predict what they would do, but once we had made up our minds we *expected* important results. However, as all funding agencies should note, we made no attempt to specify what they might be or even the general areas in which they might come. Goals selected by consensus rarely surprise.

Of the 26 groups that were running at the initiative's close in 1990, perhaps 14 made transformative discoveries: that is, they did radically change the way we think, and several succeeded in achieving important scientific objectives that their peers had thought were impossible or irrelevant. Ken Seddon's and Martyn Poliakoff's Venture Research contributions, for example, have transformed the field of green chemistry, and might be one of the most important developments in industrial chemistry in the last 50 years. If other developments are included, they could lead to an industrial value of more than a billion pounds in the next decade—all for a total initial investment of some £15 million.

However, all that lay in the future. Meanwhile, having established mutually trusting relationships, we worked hard to maintain them. We did this by generally keeping in touch with progress and by periodic visits to their labs. We also arranged annual two-day meetings at our headquarters in London to which all Venture Researchers were invited. It took a few years to work out a viable format for these meetings, covering the entire spectrum of research as we did. We knew that we had got it about right when the attendance, which was not compulsory, of the more than 100 participating scientists reached close to 100%. Indeed, many researchers told us that despite the plethora of conferences nowadays, ours was a "must go." One of their attractions was that they were a festival of *science*— specific disciplines were rarely mentioned, and we hope that Pasteur would have been at home. Our format was to allocate roughly equal time to presentations and discussions, to abolish the concept of the stupid question, and to have extended breaks between each day's few presentations. As all our researchers were pioneers, they seemed to welcome the opportunity to share their trials with fellow crusaders. Their discussions also led to several informal and fruitful collaborations that we usually funded separately.

We also held frequent workshops on general topics arising naturally from our program. They usually involved about five of our research teams and about a dozen or so BP scientists, and lasted two days. Technology transfer has long been fashionable, but it is too passive a term. Know-how cannot be transferred like a baton in a relay race. There must be a meeting of minds, and we found that the most effective transfers (a two-way process) occur when people are stimulated to see their own problems in a new light. There are no intellectual obstacles in these circumstances as one might reasonably claim that any new ideas were one's own. Workshops with such titles as "coherent systems," "theoretical engineering," and "organization and behavior in biology" arose ad hoc and were simply one-off meetings on a broad and topical subject. One of those workshops happened to coincide with the breaking news of Fleischmann and Pons's cold fusion "discovery," and we dedicated a couple of hours to debating the published claims (Fleischmann et al. 1989). We very quickly concluded that as far as nuclear fusion was concerned, they were nonsense. Deciding to stick my neck out, I reported our opinion the next day to a meeting of senior BP staff, some of whom were not very pleased as they were hoping to jump on the bandwagons that many shortly expected would roll.

The workshops on computing science were different from the others, and we had about 10 workshops on that subject. This was because we had the good fortune to have Edsger W. Dijkstra's participation in our program.[5] Tony Hoare, professor of computation at Oxford University, said of him in 1986:

> Dijkstra occupies in Computing Science the same position that Einstein did in Physics in the 1920s. Any physicist who did not know his name would hardly be accepted as a physicist. The work of Dijkstra in solving the problems of software engineering is as important as the work of Einstein in the generation of power or the manufacture of bombs. But in contrast to Einstein, Dijkstra is keen to pursue his research into the development phase and right through to application. (Private communication)

Dijkstra was indeed one of the most profound thinkers I have met, and a consummately skilled practitioner in the use of Occam's razor. His Venture Research, which extended over nine years in collaboration with either Netty van Gasteren or Lincoln Wallen, was entitled variously "The taming of complexity" or "The streamlining of the mathematical argument." It might also have been informally called "To erect barriers to the introduction of complexity." One of his first remarks was that the usual name given to his field, *computer science* (as opposed to *computing* science), was a meaningless term because it was like referring to surgery as "knife science." Unfortunately, his rigorous concentration on fundamental

---

5   At Eindhoven University of Technology, and later, at the University of Texas at Austin.

principles, and his sometimes tactless remarks to those who might not be inclined to agree,[6] annoyed some BP scientists and engineers. They did not appreciate Dijkstra telling them, as in this trivial example, that zero was a number; they knew that very well, they said. But when asked to name the lowest integer, how many would reply with "zero"? Education from primary level onward often treats zero as a special case rather than merely as an even number with certain properties. In computing science, failure to recognize zero's status can lead to serious problems. BP was divided, therefore, into people who either hated or admired him. Consequently, technology transfer was difficult because some people did not agree that he had anything to transfer! However, our involvement with Dijkstra persuaded several eminent scientists to participate in our Venture Research program, which in turn eventually led to BP creating an in-house Information Technology Research Unit. Not surprisingly, BP Exploration was a major user of computers. At that time, the global demand for data processing in oil exploration was larger than in any other field, exceeding even military and defense requirements. Nevertheless, this was a major departure for BP, and the ITRU was subsequently very influential. Its first director, Tom Hill, acknowledged that much of its success would not have been possible without access to Venture Research's academic connections, which themselves were made possible only by virtue of our association with Dijkstra.

In 1990, our stock within BP generally could hardly have been higher, and we were also beginning to establish a wider-based reputation for excellence. Nevertheless, in March that year, a few gray men decided that our magnificent initiative should end.[7] Fortunately, managing directors intervened to allow our 10th-anniversary conference to go ahead in the summer, to which we had invited about 250 people. The conference was our most successful, and many scientists (including many senior BP research staff) expressed bewilderment that such a huge and enlightened company as BP could not find a niche to accommodate such a tiny jewel in its enormous crown.

---

6   Dijkstra loved to lampoon pompous managers, and sometimes delivered his scathing remarks in the guise of written pronouncements from the chairman of his fictitious company, Mathematics Inc. For example, he once had that mischievous chairman make the following remark: "With Mathematics Inc., by the way, we are in trouble. No matter how corrupt our commercial practice, no matter how fraudulent our scientific activities, the world around us seems to beat us up. In these competitive times, it is bloody hard even to catch up with reality!" (Dijkstra 1982, 204)

7   See my *Pioneering Research* for a review of that story. Briefly, we became vulnerable following two retirements. Robert Malpas succeeded Jack Birks in 1982 as our contact managing director. He did so with great enthusiasm. When he retired in 1988, I was unable to prevent the transfer of the Venture Research Unit into the protective custody of the research director, John Cadogan. Sir Peter Walters, BP's chairman and chief executive, retired in 1990. He had been perhaps our staunchest ally for over eight years. Within days of his going, the Unit was closed down.

Those who set up TR initiatives should therefore ensure that the arrangements are consistent with long-term stability. But they should not be immune from assessments even on relatively short timescales. One of the inherent advantages of TR is that one can judge its quality long before projects come to fruition. Within a few years, it should be possible to differentiate TR from mainstream research by the exceptional ability of its scientists and their open-ended approaches. Indeed, one of the reasons we took the high-profile course of holding regular conferences in BP's head office in London was to allow staff to see for themselves the inspirational qualities of the scientists involved. It was a high-risk strategy, of course. Had they been run-of-the-mill, we would not have lasted long. But it was effective, and the conferences played an essential role in Venture Research's survival for over 10 years.

However, TR initiatives must be protected from the inevitable attacks by vested interests. Such attacks are as old as the hills and have determined scientific progress since Aristotle's time. Nevertheless, these initiatives will need sustained and committed high-level protection because some officials will believe that successful TR initiatives would be living proof of the serious flaws in their policies. In these circumstances, they might be tempted only to offer "every assistance short of actual help," as the saying goes. TR staff must do everything they can to soften any criticism implied by their very roles, but as my experience at BP suggests, the more successful they are, the bigger the threat to their existence would become. My recommendation is that the best protection would come from basing them on the broadest possible infrastructure. Before 1970 or so, that infrastructure would have been civilized society itself, when it was arranged, by accident or otherwise, that most researchers had access to modest funds more or less on demand. We have moved on, of course, but consideration might be given to setting up TR initiatives as public-private partnerships. In addition to the advantages mentioned in Chapter 3, it could mean that these initiatives are not as susceptible to ex cathedra decisions on such issues as closure from one of the partners. The arrangement might also act to soften any implied criticism. Such collectively funded high-profile initiatives would be so obviously different from anything that had gone before that comparison could be avoided.

As ever, getting the right people will be vital to success. My suggestion would be that the TR director should be selected in consultation with senior scientists and then invited to take whatever action he or she deems appropriate. The director would appoint others to work alongside as required, subject to the approval of an appropriate board. Staff should be prepared to serve for five years or preferably longer. This would help build mutual trust—there would be no point in offering to back researchers through thick and thin if you will not be around to keep that promise. It also concentrates the mind if your term of office is sufficiently long to allow you to reap the credits or whirlwinds as they materialize. Bureaucracy is, of course, the archenemy of research, and the director should be free to reduce it to the lowest level deemed compatible with efficient operations. Indeed, TR staff should fiercely resist trends that might lead to increased bureaucracy. As their program grows, the director may wish eventually to consider ways of maintaining personal and direct involvements in the face of increasing demands. A large national program might be split into a number of regional programs, for example, with an overall director to ensure that regional entities operated along similar lines.

The likely costs can be estimated using back-of-the-envelope figures. Let us assume that there were about 300 transformative researchers—the extended membership of the Planck Club—during the twentieth century. Let us adopt a rule of two by which we increase or decrease cost estimates by a factor of two, whichever is the most pessimistic. Allowing for inefficiencies, therefore, let us increase the target number of transformative researchers we must find to 600—that is, six a year on average over the century. This is a global estimate, but for a TR initiative in a large country such as the United States, then, according to our rule of two, we assume that *all* the new members might have to be found in that country since it is home to about half of all R&D. If we also assume that the searches will be about 50% efficient, which Venture Research experience indicates would be about right, it would mean that a US TR initiative should find some 12 transformative researchers a year. (For comparison, a maximum of nine scientists can win the Nobel Prizes each year). TR is the cheapest research there is, as it is heavy on intellectual requirements but relatively light on resources. For Venture Research in the late 1980s–early 1990s operating in Europe and the United States, the average cost per project was less than £100,000 a year, including all academic and industrial overheads. Costs have gone up since then, so for our present purposes we might double them to, say, £200,000 or $400,000 a year per project on average.

Transformative researchers should be supported initially for three years. Our experience indicates that about half of them would require a second three-year term; and half of those, a third term of support. Very few projects should run for more than, say, nine years. Those leaving the TR scheme either would have succeeded and been transferred to other programs created for them—that is, their research would actually have been transformative—or the scientists agree that they had probably failed in their Herculean quests. However, these average figures are quoted for guidance; there should in fact be no hard-and-fast rules on the length of support. Remember Planck!

So, the back of the envelope should now be ready. If x is the number of new research projects selected each year, then according to these rough rules, the number of projects being supported after three complete three-year cycles, or nine years, would be

$$3x + \frac{3x}{2} + \frac{3x}{4} = 5.25x$$

This sum should also be the steady total thereafter. As we have chosen *x* to be 12, after nine years, therefore, a TR research budget (i.e., excluding overheads such as the initiative's administrative costs) would be some $25 million a year. If it turns out to be significantly more than that, the initiative would be tackling a different problem than TR. After the first nine years, the TR initiative would have backed some 108 projects, of which according to our experience about 54 should eventually turn out to be transformative in some way.

A TR budget for a smaller country—say, the United Kingdom—should be about half that of the United States, or $12.5 million per annum. The Venture Research budget in our final year of operation (1990) was some $5 million, two-thirds of which we spent in Britain. As we had been operating for 10 years, it is possible that we had identified most of the researchers in Britain looking for potentially transformative research support at that time.

7   The Venture Research Harvest

"Old recipe" for Venture Research:

- Take a senior scientist eager to take vicarious pleasure in other people's research.
- Give that scientist a modest bag of money.
- Take hungry researchers plotting revolution.
- Prepare mutual trust.
- Cut out all bureaucracy.
- Season generously with freedom and encouragement.
- Stir, taste, and contemplate.

British Petroleum set up the Venture Research Unit in 1980 and closed it down in 1990. I was the head of the Unit throughout that time. Twenty-six projects were running at the time of closure, but as we awarded grants for three years, some coasted on after 1990. They all finally expired in 1993. However, for the final phase after 1990, researchers were on their own; that is, they were not provided with the usual Venture Research services described in Chapter 6 and summarized above in the "old recipe." Several teams could not get further funding as they were immediately and prematurely exposed to the full rigors of peer preview. These teams might have made substantial progress, but it was nevertheless insufficient either to create new mainstreams or to convince funding agencies that the new lines were worth pursuing.

When the Unit was first set up, we did not have a viable strategy. Hence, we were cautious, and much of our energies went into discussions about what we should do. Indeed, it took about four or five years before we were confident that our strategy would be effective. As a result, the Unit's expenditure during the first five years was much less than half the total for the decade. The research supported during that first period was certainly of high quality as otherwise our BP board would not have approved it. However, its coefficient of adventurousness, so to speak, was generally less than that of the latter five years because we then knew precisely what we were looking for.

The reasons for selecting each participating team and, where possible, some of their subsequent achievements, are summarized in the following pages. All except one had been unsuccessful with the conventional funding agencies. Indeed, we were usually the agency of last resort. The total cost of the program for the period 1980–1990 was approximately £15 million, including all BP and university overheads. BP honored its financial commitments during 1990–1993 at a further total cost of about £5 million. The average cost of the nine projects approved in 1989, the last full year of normal operations, was £287,671 ($460,274) for a three-year program, including university overheads.

The review covers all the 26 projects running at our closure. My assessment is that almost all of them were scientifically successful—that is, the Venture Researchers generally achieved objectives that their peers had thought either impossible or irrelevant, and went on to produce many significant publications. At least 14 proved transformative, and these are listed in **Table 13**. (Poliakoff and Seddon were completely separate projects, but their contributions to green chemistry are now usually taken together.)

Our experience therefore confirms the huge harvest awaiting those who would find ways of giving individuality completely free rein. Venture Research was global in scope, but our budget—determined entirely by the rate at which potentially transformative ideas could be brought forward—never exceeded 0.2% of the UK expenditure on academic research, and, of course, was therefore only a minuscule fraction of the global total. Individuality did run more or less free before about 1970, but freedom for everyone is impossible now.

Table 13 would seem to prove conclusively that for radical researchers at least, a Venture Research-style initiative does indeed offer a viable alternative to the environment prior to about 1970. However, although the program's results have been published in the literature, they are widely dispersed. There would be little or nothing to indicate a common source, for example. I have therefore gathered as much of the information as possible together for the first time—it is perhaps 90–95% complete—and presented it as a story. Although we have been unable to raise the funds to replicate the program (as of 2007), I hope that it will either help our quest, or inspire others to reproduce it themselves.

Table 13: Some Transformative Discoveries by Venture Researchers

VR 1	Mike Bennett and Pat Heslop-Harrison	Discovered a new pathway for evolution and genetic control
VR 3	Terry Clark	Pioneered the study of macroscopic quantum objects
VR 4	Stan Clough and Tony Horsewill	Solved the quantum–classical transition problem by developing new relativity and quantum theories
VR 7	Steve Davies	Developed small artificial enzymes for efficient chiral selection
VR 10	Nigel Franks, Jean-Louis Deneubourg, Simon Goss, and Chris Tofts	Quantified the rules describing distributed intelligence in animals
VR 12	Herbert Huppert and Steve Sparks	Pioneered the new field of geologic fluid mechanics
VR 14	Jeff Kimble	Pioneered squeezed states of light
VR 15	Graham Parkhouse	Derived a novel theory of engineering design relating performance to shapes and materials
VR 16	Alan Paton, Eunice Allan, and Anne Glover	Discovered a new symbiosis between plants and bacteria
VR 19	Martyn Poliakoff	Transformed green chemistry
VR 22	Ken Seddon	Transformed green chemistry
VR 23	Colin Self	Demonstrated that antibodies in vivo can be activated by light
VR 24	Gene Stanley and José Teixeira	Discovered a new liquid–liquid phase transition in water that accounts for many of water's anomalous properties
VR 25	Harry Swinney, Werner Horsthemke, Patrick De Kepper, Jean-Claude Roux, and Jacques Boissonade	Developed the first laboratory chemical reactors to yield sustained spatial patterns—an essential precursor for the study of multidimensional chemistry

The following is an index of my summaries. Groups are listed alphabetically in their present (2007) university locations:

VR 1   Mike Bennett, Kew Gardens; and Pat Heslop-Harrison, Leicester: "The Nature and Significance of Higher Order Genome Structure," 1986–1992

VR 2   Paul Broda, University of Manchester Institute of Science and Technology: "Biodegradation of Lignin," 1981–1991

VR 3   Terry Clark, Sussex: "Macroscopic Quantum Objects," 1985–1991

VR 4   Stan Clough and Tony Horsewill, Nottingham: "Coherence in Condensed Matter Dynamics," 1989–1992

VR 5   David Cooper, Liverpool; and Joe Gerratt, Bristol: "Understanding the Electronic Structure of Solids," 1990–1993

VR 6   Adam Curtis and Chris Wilkinson, Glasgow: "Understanding Nerve Circuits," 1985–1991

VR 7   Steve Davies, Oxford: "Understanding Molecular Architecture," 1985–1991

VR 8   Edsger W. Dijkstra, Texas at Austin; Netty van Gasteren, Eindhoven; and Lincoln Wallen, Oxford: "The Taming of Complexity," 1981–1990

VR 9   Peter Edwards and David Logan, Oxford: "Metallic, Non-Metallic, and Exotic States of Matter," 1987–1992

VR 10   Nigel Franks, Bristol; and Jean-Louis Deneubourg and Simon Goss, Université Libre de Bruxelles: "Collective Problem-Solving," 1990–1993

VR 11   Dudley Herschbach, Harvard: "Dimensional Scaling as a New Calculus for Electronic Structure," 1988–1991

VR 12   Herbert Huppert, Cambridge; and Steve Sparks, Bristol: "Multi-Phase Flows in Dense Media," 1983–1992

VR 13   Andrew Keller, Ted Atkins, and Peter Barham, Bristol: "Polymer Transition Dynamics," 1984–1990

VR 14   Jeff Kimble, California Institute of Technology: "Quantum Dynamics of Optical Systems," 1983–1992

VR 15   Graham Parkhouse, Parkhouse Consultants: "An Integrated Approach to the Theory of Structures," 1986–1992

VR 16   Alan Paton, Eunice Allan, and Anne Glover, Aberdeen: "Induction of Novel Symbioses between Bacteria and Higher Organisms," 1982–1992

VR 17   John Pendry, Imperial College of Science and Technology: "Transport in Disordered Systems," 1982–1991

VR 18   John Polanyi, Toronto: "Surface Aligned Photochemistry," 1983–1991

VR 19   Martyn Poliakoff, Nottingham: "Supercritical Fluids: An Environment for Reaction Chemistry," 1988–1991

VR 20   Alan Rayner and John Beeching, Bath: "Towards Understanding Multicellularity," 1987–1990

VR 21   Ian Ross, University of California at Santa Barbara: "Cytoplasmic Control of Nuclear Behaviour," 1988–1991

VR 22   Ken Seddon, The Queen's University of Belfast: "Chemistry and Physics in Ionic Liquids," 1988–1991

VR 23   Colin Self, Newcastle: "Biological Instability," 1990–1993

VR 24   Gene Stanley, Boston; and José Teixeira, Laboratoire Léon Brillouin, CEA-CNRS, France: "Water in Confined Geometries," 1990–1993

VR 25   Harry Swinney, Texas at Austin; Werner Horsthemke, Southern Methodist University, Dallas; and Patrick De Kepper, Jean-Claude Roux, and Jacques Boissonade, Centre de Recherche Paul Pascal, Bordeaux: "Self-Organisation in Non-Linear Chemical Systems," 1985–1991

VR 26   Robin Tucker, David Hartley, and Desmond Johnston, Lancaster: "Geometrodynamics," 1990–1993

VR 1. Mike Bennett, Kew Gardens; and Pat Heslop-Harrison, Leicester: "The Nature and Significance of Higher Order Genome Structure," 1986–1992

I first met Mike Bennett and Pat Heslop-Harrison at the Plant Breeding Institute in Cambridge following an invitation from the director, Peter Day. After the introductions, Day explained how embarrassed he was that he could not support the group's proposed research because it lay outside the Institute's remit. Indeed, as they wanted to carry out a coherent study of the hierarchical structure of genomes in general—that is, those of animals, humans, and plants—it would have required the concerted approval of three autonomous UK research councils, a feat that even for noncontroversial research would have bordered on the impossible. To make matters worse, it was generally thought in 1986 that cell nuclei were amorphous; that is, their constituents were immersed in what was often called a "nuclear soup," or alternatively, "nuclear spaghetti." Conventional wisdom therefore insisted that there was no structure to study, so their proposals were consistently turned down. As one of Nature's ambassadors, its simplicity appealed to me immediately, and I could not see it supported quickly enough.

The group was the first to demonstrate that genes and the chromosomes on which the genes are carried are arranged nonrandomly within three-dimensional structures. They showed that a chromosome's three-dimensional position is correlated with and regulates the development of cells, organisms, and species. They also discovered some of the rules governing the regulation and behavior of suites of genes as genomes evolve and hybridize with time. These changes can be substantial, and give heritable changes independently of mutations. Thus, the group would seem to have discovered a new pathway for evolution and for genetic or developmental control.

Bennett and Heslop-Harrison have therefore opened up a new dimension, quite literally, in our understanding of biology. However, despite this prodigious success, they were unable to obtain further support when their Venture Research funding came to an end in 1992. Their problem seems to be that their work is not regarded as relevant, in the current jargon. Additionally, it is not based on such fashionable organisms as *Arabidopsis* or rice, and would not necessarily lead to new genome products. Their proposals are met with nitpicking quibbles and are considered "too broad" in scope. Thus their novel and proven approach is being ignored. If it could be continued, it would almost certainly lead to a new understanding of disease, of the dynamic structure of the genome, and of such essential topics as gene expression.

General References
1. J. S. Heslop-Harrison and M. D. Bennett, "Nuclear Architecture in Plants," *Trends in Genetics* 6 (1990): 401–405.
2. J. S. Heslop-Harrison, "Comparative Genome Organization in Plants: From Sequence and Markers to Chromatin and Chromosomes," *Plant Cell* 12, no. 5 (2000): 617–635.

VR 2. Paul Broda, University of Manchester Institute of Science and Technology (UMIST): "Biodegradation of Lignin," 1981–1991

Lignin is the major structural component of all plant material. It is unique among the abundant natural polymers in being formed by chemical polymerization rather than biosynthesis. It is also insoluble, three-dimensional, and non-chiral with a variety of bond types in random order. Moreover, it combines in Nature with cellulose and hemicellulose to form lignocellulose, the world's principal renewable resource. In 1980, Paul Broda was about to move to UMIST to take up a new appointment, a time when researchers are often in an adventurous frame of mind. He therefore proposed to tackle the apparently intractable problem of lignin biodegradation, arguing that there could be substantial environmental benefits in such fields as animal nutrition or pulp-and-paper production if its biodegradation were better understood. The only organisms known to break down lignin completely in pure culture are the white rot fungi, *Phanerochaete chrysosporium*, for example, so Broda ambitiously proposed using the new techniques of molecular genetics to study both lignin degradation and white rot fungi. In parallel, his group also proposed studying actinomycete bacteria (which are also filamentous) to establish whether they, too, can degrade lignin.

We were delighted to support such a broadly based study of an important topic, and in a fine example of what can be done by public-private partnerships (see Chapter 4). Broda was also able to persuade the Agricultural and Food Research Council (which no longer exists) to support the work in parallel with Venture Research, thereby doubling his money, so to speak. We were happy to have catalyzed the amplification.

Three principal themes emerged from Broda's pioneering approach:

1. His group developed methods for DNA and messenger RNA isolation from fungi that became widely adapted. Thus, they isolated clones carrying genes specifically and strongly expressed during lignin degradation. Some of these were lignin–peroxidase-related sequences differentially expressed under different physiological conditions. They also applied the newly discovered technique of polymerase chain reaction to analyze the differential expression of closely related lignin peroxidase and cellulase genes.
2. His group was the first to use fungi in restriction fragment length polymorphisms for genetic mapping, and for strain improvement.
3. His group was the first to study hemicellulose degradation by *Phanerochaete chrysosporium*.

His group also showed that the actinomycete bacteria can at least solubilize lignin if they cannot degrade it. They characterized enzymes that break down the hemicellulose component (xylanases), and developed substrates for more rigorous assays of lignin and hemicellulose degradation.

Broda's pioneering work continued until 1995 and provided a rigorous basis for the future dissection of this complex, fundamental, and technologically important process. However, his Venture Research support ended in 1988 when BP deemed that support for his work should be transferred to BP Nutrition, then a member of the BP group. The rationale was that as livestock animals are unable to extract the nutrients that lignin encapsulates, Broda's approach could considerably improve the nutritional value of traditional animal feeds, then one of BP Nutrition's major businesses, by improving the accessibility of previously entrapped nutrients. *This thesis remains valid today*; the nutritional value of such feedstuff can indeed be improved by reduction of lignin content. Moreover, the bovine spongiform encephalopathy saga has highlighted the dangers of inappropriate and unnatural feedstuff. Unfortunately, the high initial costs of enzyme production mean that an enzyme-supplemented composting-type treatment has been deemed economically unviable. The alternative possibility of using genetically engineered organisms now seems unacceptable to the public, and has not been rigorously explored.

The pulp-and-paper industry is one of the most energy-intensive and historically polluting, partly because of the need to remove and/or bleach the lignin component from wood pulp. Chlorine and its compounds are no longer used for bleaching, having been replaced by molecular oxygen. Although there has been some development of enzyme processes for pulp bleach enhancement, de-inking, pitch removal and fibrillation, microbial technologies for decolorization of effluents, and biopulping to reduce the use of chemicals and energy, Broda's elegant techniques have had little impact largely because of the initially high capital costs. However, in this and the animals feed case, it would seem that economies of scale have been largely ignored.

General References
1. B. S. Hartley, P. M. A. Broda, and P. J. Senior, eds., *Technology in the 1990s: Utilisation of Lignocellulose Wastes* (London: The Royal Society) 1987.
2. P. M. A. Broda, P. R. G. Birch, P. R. Brooks, and P. F. G. Sims, "Lignocellulose Degradation by Phanerochaete Chrysosporium: Gene Families and Gene Expression for a Complex Process," *Molecular Microbiology* 19, no. 5 (1996): 923–932.

VR 3. Terry Clark, Sussex: "Macroscopic Quantum Objects," 1985–1991

The very title of his proposed Venture Research raised serious problems for him among the peer-review establishment simply because quantum-mechanical behavior had never been associated with *macroscopic* objects. However, Clark's reasoning opened our eyes to a wide range of new and unexplored science. As he explained, when quantum mechanics was first formulated by Schrödinger, Heisenberg, and others in the 1920s, it was specifically designed to predict the properties and behavior of microscopic objects—electrons, atoms, molecules, and light particles (photons)— and there was an implicit belief that this behavior would be observed only in the microscopic domain. Nevertheless, it was soon realized by Bose and

Einstein that particles carrying zero or integer quantum spin followed a special statistics, subsequently known as *Bose–Einstein statistics*. Furthermore, they demonstrated that below a certain critical temperature a collection (a gas) of these particles—"bosons"—could undergo a transition to a single quantum state described by a simple, macroscopic parameter, and usually referred to as a *condensate*. It was already known that superconducting behavior occurred in certain metallic materials at sufficiently low temperatures, and that it closely resembled the dissipationless behavior of electrons bound in atoms and molecules. In the 1930s superconductivity was joined by a new stablemate—super-fluidity—where the dissipationless flow was that of chargeless helium-4 atoms.

It might have been realized then, therefore, that quantum mechanics was not restricted to the microscopic domain. However, it was argued that a macroscopic condensate is not a quantum object. It took many developments, in both theory and experimental technique, for the stage to be set for the construction and observation of macroscopic quantum objects. Of outstanding importance in this respect was the prediction by Brian Josephson in the early 1960s of "weak superconductivity" in the coherent tunneling of electron pairs between two bulk pieces of superconductor—the famous Josephson effect. These devices are now called *weak links*. This was followed in the late 1960s by the invention of the SQUID (superconducting quantum interference device) ring, which eventually transformed the argument about the possibility, or otherwise, of macroscopic quantum objects. In its simplest form the SQUID comprises a thick superconducting ring enclosing a Josephson weak link.

Clark had realized that the SQUID ring *is* a macroscopic quantum object, although for many years it was generally regarded as quasi-classical—a kind of halfway house between the classical (Newtonian) and quantum worlds. This typically one-centimeter-diameter object has the remarkable property that its energy levels can be changed by the application of extremely small amounts of magnetic flux. Furthermore, shifts in energy levels were periodic in this flux. The flux dependence, and the periodicity, arises because of the remarkable potential (energy) dependence of the SQUID ring on external flux (a parabola with a cosine modulation). When an external flux is applied, the potential rolls one way or the other, as he put it, depending on the direction (sign) of the flux, shifting the ring energy levels as it goes.

Clark went on to tell us that this simple picture constituted no less than a paradigm shift and opened the way to studying macroscopic quantum objects for the first time. All that was required to probe the quantum evolution of a SQUID ring was to couple the ring to an external, classical inductive circuit. In practice this would take the form of a linear parallel inductive-capacitive resonant circuit, usually resonating at radio frequencies, and which, of course, is most definitely macroscopic. The quantum evolution of the SQUID ring can then be followed through shifts in the resonant frequency of the classical part of the coupled inductive system. In order to excite its resonance, Clark's team had created a technique whereby the excitation came from the intrinsic noise in the low-temperature (low-Kelvin-range) classical circuitry coupled to the SQUID ring.

Despite the exhilarating prospects Clark had outlined, his group was about to lose its SERC support (see **Poster 11**). As they were all on "soft" money (Clark did not have tenure), the group not only was about to be disbanded but would also be unemployed. We were delighted to step into the breach.

As their work continued, it revealed a further twist to the interaction of the quantum and the classical. The dynamical inductance of the SQUID ring is a nonlinear (and periodic) function of the applied flux. Thus, the observed inductance of the coupled resonator is also nonlinear, and periodic, in this applied flux. Thus, a macroscopic quantum object, a SQUID ring, can not only manifest its quantum evolution in a classical, macroscopic circuit structure but can also induce nonlinear, and often extreme, behavior in that structure. Quantum mechanics has, of course, been widely studied using optical systems, but it turns out that Clark's SQUID-based system is much more efficient at creating nonlinear effects, and hence may be a more sensitive probe of quantum mechanics. All this, and much more, was revealed in the course of their Venture Research work, as the following short list indicates:

1. M. J. Everitt, T. D. Clark, P. B. Stiffell, et al., "Superconducting Analogs of Quantum Optical Phenomena: Macroscopic Quantum Superpositions and Squeezing in a Superconducting Quantum Interference Device Ring," *Physical Review A* 69, no. 4 (2004): 3804; idem, "Quantum Downconversion and Multipartite Entanglement via a Mesoscopic Superconducting Quantum Interference Device Ring," *Physical Review B* 72, no. 1 (2005): 4508.
2. J. F. Ralph, T. D. Clark, M. J. Everitt, et al., "Nonlinear Back-Reaction in a Quantum Mechanical SQUID," *Physical Review B* 64, no. 18 (2001): 1805.
3. T. D. Clark, J. F. Ralph, M. J. Everitt, et al., "Coherent Evolution and Quantum Transitions in a Two Level Model of a SQUID Ring," *Annals of Physics* 268, no. 1 (1998): 1–30.

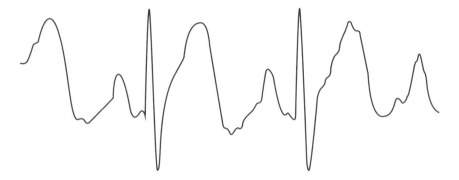

Figure 13: The author's electrocardiogram taken in 2001 with handheld equipment from Terry Clark's lab.

4. R. Whiteman, T. D. Clark, R. J. Prance, et al., "Microwave-Induced Transitions in a Superconducting Quantum Interference Device," *Journal of Modern Optics* 45, no. 6 (1998): 1175–1184.
5. T. P. Spiller, T. D. Clark, R. J. Prance, et al., "The Adiabatic Monitoring of Quantum Objects," *Physics Letters A* 170, no. 4 (1992): 273–279.
6. J. F. Ralph, T. D. Clark, R. J. Prance, H. Prance, "Self-Capacitance of the Superconducting Condensate," *Physica B* 226, no. 4 (1996): 355–362.

Clark's group also developed new electronic techniques for characterizing weak electronic signals with high spatial resolution. These techniques, developed initially for their SQUID studies, have proved to have a wide range of applications, particularly in biology. In effect, these techniques can noninvasively measure voltages to exceptionally high precision in time and space. Thus, their handheld heart monitor can immediately give high-resolution electrocardiograms of the heart's performance throughout its cycle. A scan of the author's heart using Clark's equipment is shown in **Figure 13**.

The group developed devices for scanning other organs, including the brain, and for scanning electronic signals from living cells. However,

---

1   The funding problems at British universities are well known, but those at the University of Sussex have attracted wide publicity. The Physics Department has been virtually closed down (Clark's group moved to Engineering, for example), and the Chemistry Department is on the verge of closure. Sir Harry Kroto, the University's most recent Nobel Laureate (Chemistry), has moved to the University of Florida. Such an environment is not conducive to contemplative research.

Clark is not a businessman, and he has always insisted that access to their group's impressive technology would be conditional on support being provided for his basic research on quantum systems. Had Venture Research continued, this would not have been a problem,[1] and his valuable technology could by now be well on the way to full commercial exploitation. Progress has been made, but is but a shadow of what it could have been.

That comment also applies to their science. The size of his quantum objects, for example, creates intriguing possibilities. By modulating the potential of quantum-mechanical SQUID rings, it might be possible to generate energy bands extending over time rather than space—that is, *time bands* rather than the more familiar spatially periodic energy bands in crystalline structures in condensed-matter systems. This could be a truly revolutionary discovery that would open up a vast new range of physics. Furthermore, bearing in mind that high-energy physicists probing the QED of nuclear systems are obliged to use high-energy and expensive γ rays (i.e., photons with wavelengths comparable to those nuclear dimensions), similar experiments (Widom and Clark 1982) could be performed with the QED inherent in SQUID rings using low-energy and cheap microwaves.

VR 4. Stan Clough and Tony Horsewill, Nottingham: "Coherence in Condensed Matter Dynamics," 1989–1992

Conceptual problems and controversy have dominated the history of quantum mechanics. In the case of systems transformed by rotation, for example, an isolated ammonium molecule cannot be distinguished from itself after being rotated through 360°; that is, we cannot count rotations in quantum mechanics. In classical mechanics, this is a trivial matter. The hands of a grandfather clock, say, look identical after 12 hours, but the weight will have fallen. The difference arises because quantum theory has to account for quantization but in so doing makes it impossible to describe classical phenomena. The team's ambitious goal, therefore, was to reconcile these apparently essential incompatibilities. Conventional quantum mechanics describes microscopic systems, and as objects in the everyday world are composed of assemblies of such systems, one theory should describe both. However, quantum mechanics has had little success in describing classical behavior. Since theoretical understanding in any field strongly influences what is done and the types of question asked, failure to reconcile the two views could inhibit the discovery of new phenomena, or restrict the ways we exploit current understanding. We had no hesitation, therefore, in supporting their proposal.

In the team's view, conventional quantum theory contains features that are neither verifiable nor unverifiable, and which are in conflict, therefore, with the philosophy of scientific method. One of these is the axiom that quantum particles are identical and indistinguishable. Classically they are distinguishable, of course, and initially, their project aimed to trace this apparent paradox through the quantum-classical transition exhibited by molecular rotors in solids and then make whatever theoretical changes should turn out to be necessary. There were three possibilities:

1. Classical theory is wrong and protons should be indistinguishable.
2. Quantum and classical theories are both slightly wrong and need modifying to make indistinguishable protons become distinguishable as part of the quantum-classical transition.
3. Quantum theory is completely wrong, and protons are not indistinguishable particles.

The generally accepted view was possibility 1, but suspiciously, no classical theory with indistinguishable particles has ever been developed. Their original expectation was that possibility 2 would be correct, but astonishingly, they have now reached the radical conclusion that the correct interpretation is possibility 3.

Since their Venture Research funding ended, the team have been considering and reflecting on what they did. It would now seem that the reason for their remarkable conclusion is that science has used the wrong concepts, and has done so since the time of Archimedes and Euclid. Stan Clough has gone on to show that none of the concepts of distance, time, velocity, and particles of fixed size and shape can actually be observed.[2] Indeed, they are independent of the observer, have an unverifiable existence, and are incompatible with the relativity principle. Instead, they are observer-dependent and defined by measurement, as are range, age, Doppler shift, and the observed variable size and shape of electrons, for example.

The switch to these concepts results in a new relativity, and quantum and classical theories that cover the same ground as the ones they replace while being mutually compatible. Their foundation is Einstein's relativity principle, the most fundamental of all physical laws. This is crucially absent from conventional quantum and classical theories and, although present in Einstein's relativity, it is expressed in concepts with which it is

---

2   S. Clough, private communication (to be published).

incompatible, making the theory more complicated and mysterious than is necessary.

The broader objectives that they subsequently set themselves imposed completely new demands. It was no longer sufficient to concentrate on molecular rotation and the concept of indistinguishable particles. It became necessary to show that their new concepts deliver all the established results of relativity and quantum theory. These include the gravitational redshift, the curvature of light by the Sun, the closed geometry of the finite universe, the precession of the perihelion of the planet Mercury, the different ages of traveling twins, the existence of black holes, antimatter, and the equivalence equation: $E = mc^2$. In quantum theory it was necessary to show that the concepts are consistent with diffraction, atoms, molecules, spin, and indistinguishable particles.

Their Venture Research support ended in 1992, but after more work and a heroic amount of contemplation they finally achieved their ambitious goal in 2007. Their achievement could turn out to be a significant milestone if not a revolution in the development of physics.

During the last century relativity and quantum theory have developed something of the status of holy writ. To suggest that Einstein, Bohr, Schrödinger, Heisenberg, and many who followed them had been getting something wrong meant that the team had to go out on a very long limb. This required enormous courage. Indeed, their skepticism was generally seen as heresy rather than an essential tool of the scientific method. In this hostile climate, the team has acknowledged that the support of the heretic-friendly and, indeed, heresy-encouraging Venture Research program played an important and much appreciated role.

General References
1. S. Clough, G. J. Barker, K. J. Abed, and A. J. Horsewill, "Molecular Friction and Nuclear Magnetism," *Physical Review Letters* 60, no. 2 (1980): 136–139.
2. S. Clough, "The Wrong Side of the Quantum Tracks," *New Scientist*, March 22, 1988: 37–40.
3. S. Clough, "Resonant Broadening of NMR Line Shapes of Tunnelling Methyl Groups at Level Anticrossings," *Molecular Physics* 78, no. 4 (1993): 781–790.
4. S. Clough, "A Molecular Rotor," *Europhysics Letters* 29 (1995): 169.

VR 5. David Cooper, Liverpool; and Joe Gerratt, Bristol: "Understanding the Electronic Structure of Solids," 1990–1993

Fashion dominates science. When David Cooper, a young newly appointed lecturer, and Joe Gerratt, a senior theoretical chemist, first came to see us, we discussed some of the problems with understanding the electronic structure of molecules and solids. The essential difficulty was that although these structures are dominated by electronic interactions, attempts to understand them have for over 50 years been based on models such as molecular orbital theory or band theory, in which it is assumed that the electrons move more or less independently of each other, experiencing only the time-averaged effects of the electronic repulsions. Although these "independent particle" models have had many successes, they cannot describe some important phenomena, such as the formation of molecules from atoms or the decomposition of molecules into products, because the electron correlations play a crucial role in determining the strength of bonds, and the point at which they form, break, or rearrange. These correlations are crucial to understanding the electrical and magnetic properties of solids.

Traditional methods had relied heavily on computers that can execute vast numbers of routine calculations in the shortest possible time. However, these brute-force methods make it difficult or impossible to incorporate what is known about chemical behavior. To give a simple example, the then-current calculations on the hydrogen molecule evaluated all possible interactions between its four constituent particles—two protons and two electrons—but took little account of the obvious fact that a hydrogen molecule consists of two hydrogen atoms. Cooper and Gerratt proposed to correct this serious defect by deferring mathematical analysis on the system in question until they had injected as much knowledge on the chemistry as possible. For some time, they had been developing a spin-coupled theory that took into account the different ways that electron spins can pair together, without imposing preconceptions. Their "spin-coupled valence bond theory" aimed to be the first quantum-mechanical treatment of chemical bonding that combined useful accuracy with highly visual interpretations, and, hopefully, would yield new insight into how systems behave. They also wished to apply their ideas to solid materials for which current approaches were hopelessly inadequate. Other approaches such as band theory, while they could be adjusted to fit data, had little predictive power and had difficulty describing electrical or magnetic properties.

We were delighted to give them the support they needed. Unfortunately, although their funds were guaranteed for three years, the Venture Research team was disbanded in 1990, a few months after the start of Cooper and Gerratt's research. This decision took some of the wind from their sails as they knew that the promised personal interactions that we normally catalyzed would not take place, and perhaps more importantly, there would be no prospect of their support being renewed. They therefore modified their ambition somewhat to concentrate on refining their methodology before tackling the more serious problems posed by solids.

Together with Thorstein Thorsteinsson and Mario Raimondi, they were indeed successful in developing a new approach to valence bond calculations. It took the tangible form of a calculative package, CASVB (Thorsteinsson et al. 1997), which was essentially a series of algorithms for performing valence bond calculations, but with the important advantage that they could be embedded in the more traditional packages based on molecular orbital theory. The CASVB codes are now part of the widely used electronic structure packages such as MOLPRO and MOLCAS.

The increased interest in valence bond theory led to the publisher Elsevier inviting Cooper to edit a new volume in its Theoretical and Computational Chemistry series: *Valence Bond Theory*, published in 2002.

Sadly, Joe is now deceased.

VR 6. Adam Curtis and Chris Wilkinson, Glasgow: "Understanding Nerve Circuits," 1985–1991

We were impressed from the outset by their unusual range of expertise—Curtis is a cell biologist and Wilkinson an electrical engineer—and their determination to tackle a fashionable problem (neural networks) from a unique perspective. The electrical activity of single nerve cells was reasonably well understood, but the interactions between even small numbers of nerve cells were not. It should not be surprising, therefore, that the signal processing in the astronomically larger human nervous system and brain is one of the least understood of any complex system. However, not only are living central nervous system networks complex—they were not amenable to experimental manipulation. Their aim was to use substrates patterned using microelectronic techniques to persuade nerve cells to arrange themselves into simple specific networks of connectivity patterns, and to study these cells over long periods using noninvasive extracellular electrodes. Thus, they hoped that the proposed work would give new understanding of the grammar and syntax of nerve signal processing at the most basic level. We were delighted to be part of it.

As they were pioneers in this field, they had to work out every detail for themselves. Crucially, of course, once they removed the neural cells from their natural environment, they had to keep them alive and functioning. That meant, for example, that they had to be kept wet; they needed access to a medium containing a suitable mix of salts, food (usually glucose, but they found that hungry cells "talk" more), and growth factors if necessary; and the "guiding" surfaces for aligning the neurons had to be biocompatible. Furthermore, they had to design and construct the surface topography that might organize cells into observable connection patterns, develop methods of extracting electrical signals from the cells over long periods, and develop the techniques for growing nerve cells in culture—another largely unexplored area. Not surprisingly, therefore, they made only limited progress during their first three years of Venture Research support, but as their proposals continued to excite us, we enthusiastically renewed the funding for another three years. Unfortunately, they had still nowhere near achieved their ambitious goals when Venture Research was closed down.

After our closure, they managed to find alternative funds from the Wellcome Trust subject to the condition that henceforth they would work on vertebrate neurons. Not surprisingly, they agreed, a point that nicely illustrates how the major grant-giving bodies not only decide who should be funded but also place subtle controls on what they might do. In 2007, we can now see that although their work has taken far longer than they originally anticipated, they now seem to have achieved after some 20 years of heroic effort all they originally set out to do. Their specific results are summarized in the following publications:

1. P. Clark, P. Connolly, A. S. G. Curtis, J. A. T. Dow, and C. D. W. Wilkinson, "Cell Guidance by Ultrafine Topography in Vitro," *Journal of Cell Science* 99 (1991): 73–77.

2. A. S. G. Curtis and C. D. W Wilkinson, "Topographical Control of Cell Migration," in *Motion Analysis of Living Cells*, ed. D. R. S. D. Wessels (New York: Wiley, 1998).

3. C. Wilkinson and A. Curtis, "Networks of Living Cells," *Physics World* 12 (1999): 45–48.

4. L. Breckenridge, P. Clark, P. Connolly, et al., "Artificially Induced Nerve Cell Patterning or Real Neural Networks," in *Synthetic Microstructures in Biological Research*, ed. J. M. Schnur and M. Peckerar (New York: Plenum Press, 1992).

5. R. J. A. Wilson, L. Breckenridge, S. E. Blackshaw, et al., "Simultaneous Multisite Recordings and Stimulation of Single Isolated Leech Neurons Using Planar Extracellular Electrode Arrays," *Journal of Neuroscience Methods* 53, no. 1 (1994): 101–110.

6. C. D. W. Wilkinson, M. Riehle, M. Wood, et al., "The Use of Materials Patterned on a Nano- and Micro-Metric Scale in Cellular Engineering," *Materials Science and Engineering* C 19, no. 1 (2002): 263.

7. A. Curtis, "Materials Science: Breaking the Neural Code," *Nature* 416, no. 6878 (2002): 274–275.

VR 7. Steve Davies, Oxford: "Understanding Molecular Architecture," 1985–1991

Davies in 1985 was a young chemist whose opening remarks to us were that much of the thinking in organic chemistry was wrong. Enzymes have taken millions of years to develop into efficient and highly specific reagents for organic synthesis. Natural enzymes (average molecular weight approximately 50,000) are excellent at controlling the chirality (handedness) of the products that they make, excellent at discriminating between enantiomers (hands—left or right) of molecules, and very efficient catalysts. Synthesis of such complex molecules would clearly be impossible, but Davies had another approach. As I explained in my *To Be a Scientist*:

> He had asked himself the question of how Nature might accomplish chiral synthesis if she had all the chemicals easily to hand that he had in his laboratory, rather than having to make do, as Nature must, with whatever materials can be found dispersed in the natural environment. He seemed to have discovered that a simple organic molecule with iron at its core behaved like a universal enzyme. It worked very efficiently, but he did not understand why it did so. We were astounded, not only by the audacity of his thinking, but also because the conventional funding agencies had declined to back the substantial and exciting program on which he now wished to embark. Reasons for rejection are rarely given, but it is likely that his youth played a part; young people are rarely authorized to attack major problems nowadays. Also, he wished to concentrate on understanding the processes of molecular architecture, rather than, as is the current fashion in organic chemistry, to promise to produce a new product virtually every week. (Braben 1994)

In the first few years of the project he developed small (molecular weight approximately 500) artificial enzymes that in terms of asymmetric synthesis (i.e., of the left or right enantiomer) and molecular recognition *are better than natural enzymes*. This transformative discovery was indeed a tour de force.

Together with Davies, BP set up a small company to begin to exploit his technology, but it was quickly overtaken by BP's decision to terminate Venture Research. Davies and others set up a new company shortly afterward: Oxford Asymmetry. In 1998, it was floated on the London Stock Exchange at a value of some £200 million.

VR 8. Edsger W Dijkstra, Texas at Austin; Netty van Gasteren (1981–1987), Eindhoven; and Lincoln Wallen (1987–1990), Oxford: "The Taming of Complexity," 1981–1990

Dijkstra had an international reputation long before I first met him. However, two things about Venture Research appealed to him. British Petroleum is one of the world's biggest companies, and Dijkstra was intrigued by the enormous challenge of trying to introduce new types of thinking into BP's decision-making. He also had a research student—Netty van Gasteren—for whom he was unable to find funds to allow her to work with him. As his research proposals also appealed to us, we were delighted to have his participation.

One of the most important problems in computing today is that we do not know how to program computers to ensure that they always execute their tasks precisely as required. The concerns involved are wholly intellectual. As for the natural sciences, there are problems of assessment and scaling up if discoveries in computing science are to be successfully exploited in the marketplace. For the natural sciences, however, the processes of development have evolved over many decades, while the equivalent processes for computing are still in their infancy. At present, the precise behavior of a substantial computer program cannot be predicted before it is used in practice. The traditional approach is to take a statistical approach to the program's completion. Thus, the program is tested until most of the errors or bugs have been identified and a point of diminishing returns is reached in the search for the rest. At that stage, the prospective failure of the system in normal operation is accepted as a reasonable risk, even though there can be no indication as to whether a possible failure will be of minor or major importance. This problem has not gone away, and may indeed have grown in importance as our everyday lives—health services, customs and immigration, traffic control, and so on—are increasingly subject to the whims of computers and the lack of precision in their programmers.

Dijkstra's broad thesis was that if computer programs could be written in the form of mathematical proofs, debugging would be unnecessary because proofs must necessarily be correct. In 1980 the domain of such programs was limited to simple examples; the aim of his Venture Research was to expand that domain into areas that could be of practical importance—the design of safety-critical software, for example. In everything Dijkstra did, from his meticulous handwriting and command of the English language onward, he stressed the need for beauty and elegance. If a proof were ugly, it must be inefficient. He deplored mathematicians who punctuated their proofs with "rabbits out of a hat"—that is, with such infuriating remarks as "it can be shown that," when in fact the steps were rarely obvious and also inhibited understanding. He advocated that proofs should contain the minimum number of lines, with each step following logically and inevitably from the previous one, as if one were walking down a staircase of even pitch—see **Poster 17**, for example.

Poster 17
—

The Pythagorean Theorem: A Beautiful Proof

Edsger W. Dijkstra was inspired by the following rabbits-out-of-a-hat-free proof of the Pythagorean theorem when he was 12 (Dijkstra 1982, 175).

The two external squares diagrammed below have the same area; therefore, they are equal.

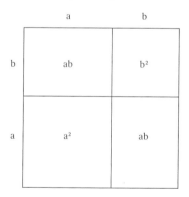 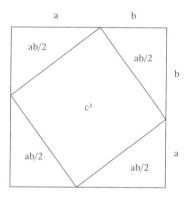

Area of left square = $a^2 + b^2 + 2ab$
Area of right square = $4ab/2 + c^2 = 2ab + c^2$
Therefore: $a^2 + b^2 + 2ab = 2ab + c^2$

Hence, by subtracting, we find that: $a^2 + b^2 = c^2$.

Formally, their objective was to improve, by an order of magnitude, on the state of the art of presentation and systematic design of algorithms and mathematical proofs. This terse description accurately describes what they set out to do, but it may not allow its significance to be readily appreciated. Even in the 1980s, there were millions of computers, and they almost all worked well for almost all of the time. However, few people trusted them, a situation that may not have changed much, but the problems stemmed mainly from the software rather than hardware. As for hardware, industrialists soon saw the need to reproduce on industrial scales laboratory conditions that do not allow one detectable speck of defect-causing dust, for example, so that hardware is virtually fault-free. Such equivalent processes for software did not (and still do not) exist, and intellectual dust abounds. There were (and still are) some large and widely used programs in which hundreds of new errors are discovered each month, and even when they are corrected, it is sometimes the case that each correction introduces more than one new error.

Mathematics is the key to reliable programming—not so much on mathematical results, but on the processes by which mathematics is done. The Stanford mathematician John McCarthy, in a seminal paper presciently published as a young man, concluded:

> It is reasonable to hope that the relationship between computation and mathematical logic will be as fruitful in the next century as that between analysis and physics in the last. The development of this relationship demands a concern for both applications and for mathematical elegance. (McCarthy 1963)

Dijkstra and his colleagues worked hard on that relationship, and made substantial progress. The objective is still as important as ever, and had BP's sponsorship of Venture Research continued until the present day, it is most likely that we would still wish to be closely involved.

Unfortunately and most sadly, both Edsger and Netty are now deceased. My letter to Dijkstra, on the occasion of his thousandth EWD, is given in **Poster 18**. One of Dijkstra's famous characteristics was that he rarely published papers in the scientific literature. Instead, he produced immaculately handwritten and numbered reports—EWDs—which he photocopied and circulated to friends and colleagues, who in turn might circulate them further afield.

Poster 18
—

Letter from the Author to Edsger W. Dijkstra, March 19, 1987

The letter proceeded as follows:

> We received via Netty[3] a few days ago, the latest batch of EWDs (991–1000). One of the great pleasures in my self-appointed task of trying to help those scientists whose work takes them far from the madding crowds of their less critical colleagues is the occasional sheer joy accompanying the dawn of new understanding, especially on those rare occasions when one has been able to make a modest contribution. They can come from say an elegant argument suddenly appreciated, or from the realisation that phenomena hitherto thought to be unrelated are part of a coherent picture, and in general cause the dropping of the proverbial penny. On balance, these pleasures can more than offset the attendant problems and frustrations of taking on the scientific establishment, some of whom are not averse to saying or implying, in effect, that—I disagree with what you say and will fight to the death to prevent you from saying it!

> The pleasant duty of responding to your 1000th EWD also provides a nice opportunity for me to acknowledge our association and friendship, which are not only longer than any other I have been pleased to have with Venture Researchers and others during the last seven years, but also have been perhaps the most stimulating and exciting. It might of course be pointed out that my ignorance in the fields you so beautifully illuminate was so profound that it could hardly have been otherwise. With some embarrassment, I might concede that there was some truth in that view, but at least I can now say with pride that thanks to your patience and skill I have seen some of the light, and that I understand what you and your close colleagues are trying to do better than many of your would-be professional peers, who not only refuse to be enlightened, as you well know, but who also seem to believe that although there might be the odd dark corner, additional lighting is generally not required.

---

3   His collaborator, Netty van Gasteren.

The EWDs have played an important part in our dialogue, and I hope that your post-bag will shortly be able to provide confirmation that they are now required reading in other parts of BP too. So, congratulations from us all in achieving a notable goal with panache and humor. Long may you continue to give confusion to your enemies and inspiration to your friends!

VR 9. Peter Edwards and David Logan, Oxford: "Metallic, Non-Metallic, and Exotic States of Matter," 1987–1992

When Humphry Davy discovered lithium, sodium, and potassium early in the nineteenth century, he worried about whether these materials were actually metals because their densities were much lower than those of, say, mercury, silver, or gold. He called them *metalloids*, a classification that the scientific community was reluctant to change for 50 years. The question of whether materials are conductors or insulators persists today. If electrons are free to move within the crystal lattice, the material is a metal; whereas for insulators, the electrons are always bound. However, it is possible that all materials will be conductors (i.e., metals) if the external pressure is high enough. Liquid hydrogen, for example, is usually an insulator, but becomes metallic at pressures above approximately 2 million atmospheres. The planet Jupiter, the largest of the solar system's gas giants, is composed mainly of hydrogen and helium. These elements may become superconducting at the intense pressures of the planet's interior, and may be the source of its high magnetic field.

In their initial proposal, the team planned to use the well-known technique of confining metal atoms in rare-gas solids such as neon or argon. These host materials were previously chosen to study atomic states because they were thought to have little effect on them. However, the team suggested that the perturbations can be substantial for a wide range of conditions, and their ideas were supported by their ability to account for a number of previously unexplained data from published work on these systems. Edwards, an experimentalist, and Logan, a theoretician, went on to make the audacious prediction that certain rare-gas hosts should generate strong permanent dipole moments in their guest atoms. Thus, they ignored the conventional wisdom that according to quantum mechanics, atomic dipole moments are absolutely forbidden. Indeed, they thought that these forbidden states might be intermediaries in the generation of new materials with exotic properties. As they were free to follow their science wherever it might lead them, their overall objective was to understand the ways metallic states may be created.

They found that the atomic dipolar state, far from being forbidden, is more widespread than had been thought previously, and they have therefore opened up a new window on science. Indeed, they thought that the new dipolar state may play a substantial role in understanding the behavior of high-temperature superconductors.

Peter Edwards was elected a Fellow of the Royal Society in 1996.

VR 10. Nigel Franks, Bristol; and Jean-Louis Deneubourg and Simon Goss, Université Libre de Bruxelles: "Collective Problem-Solving," 1990–1993

When the team came to discuss their ideas with us in 1989, they explained current thinking on animal intelligence. Accordingly, humans occupy one end of a wide spectrum of abilities, exhibiting the highest levels of individual intelligence, while ants are found at the other. Individually, however, ants seemed less intelligent than any other animal, but communities of ants could be very intelligent. This is not because they are directed and organized by an intelligent individual such as a queen; her role is solely reproductive. Their intelligence stems from the bottom up, and would seem to be drawn solely from the communities in which they live. How does that come about? How is it organized? Their behavior seemed to be governed by a small number of simple rules. What are they?

The conventional view was that the behavior of ants and other social insects is preprogrammed and determined by an insect's age. Franks did not agree. According to convention, if a colony were to suffer a catastrophe, for example, that killed a large number of workers—young or old—the resultant inefficiencies could persist for years. But that is not what is observed. He had been impressed with the work done by the Prigogine school in Brussels, who took the view that control in biological systems is based on self-organization. Franks had therefore begun to collaborate with members of that group to see if the principles of self-organization, previously limited to studies of pattern formation in inorganic chemistry and developmental biology, might also be applied to the elaborate patterns of behavior seen in animal communities. Their preliminary data had led to their proposal. They were now proposing that ant colonies would provide a unique experimental model system because unlike interactions between molecules, interactions between workers can be observed directly, and unlike cellular systems or natural neural networks, insect societies can be easily and rapidly taken apart and reassembled.

Ant societies use self-organization to solve problems that are beyond the scope of their individual members. Such *distributed intelligence* is used in finding the shortest routes in a network, synchronizing activity and sorting items, and in task allocation. The collective problem-solving of ants is often massively parallel and actively uses feedback, randomness, and redundancy (see **Poster 19**). Indeed, the algorithms used by ants are often robust, reliable, and fault-tolerant. As their goal was to discover the structure of these algorithms, we suggested collaboration with a researcher (Chris Tofts) from the Calculus of Communicating Systems Laboratory at the University of Edinburgh,[4] and all parties were eager to accept. We were very happy, therefore, to fully support their heretical ideas.

Poster 19
—

Collective Problem-Solving According to Ants

The Franks, Deneubourg, and Goss algorithms allow for positive feedback, randomness, and redundancy in animal behavior. The idea that ants seem to use such algorithms was heretical at the time of their original proposal.

Let us say that an ant colony is exploiting many scattered food sources of varying quality and quantity:

- **Positive feedback** means that ants following and reinforcing the same chemical (pheromone) trail leading to a food source can produce rapid and efficient exploitation.

- **Randomness**, that is, wandering off an existing trail, may enable ants to discover alternative food sources before the first one is exhausted. (Venture Research ants?) Randomness also allows individuals to break out of suboptimal conditions.

---

4   Earlier in the decade, we had supported a three-year Venture Research program at the precursor to that Lab by Rod Burstall, Gordon Plotkin, and Robin Milner.

- **Redundancy**: Having very large numbers of very similar units performing the same tasks means that individual failures, for example, individuals becoming lost, do not cause the system to break down.

Such collective problem-solving can be reliable, flexible, fault-tolerant, and fundamentally simple.

The collaboration with Tofts turned out to be very successful, and led to the first application of that calculus to any field outside computation (Tofts and Franks 1992). In addition, despite the fact that the team soon realized that their Venture Research funding would not be renewed no matter how successful they would be, they responded magnificently to our encouragement to make full use of the freedom allowed. Thus, flying in the face of conventional wisdom, they decided from the outset to examine the hypothesis that ant behavior was organized from the bottom up; that is, ants are free to choose what to do on the basis of stimuli coming from changing task demands rather than being hardwired by age.

The astonishing result, however, was that the algorithms they derived revealed that ants have *both* individual and collective intelligence. Indeed, far from what was thought in 1989, they found that individual ants can show high levels of intelligence and altruism. Only recently, Franks and Richardson (2006) have shown that individual ants can sometimes be extraordinarily intelligent, and have published the first rigorous account of how an animal, in this case an individual ant, sets out to teach another about the location of a food source. The usual method in large ant societies is by broadcasting, through the laying of pheromone trails, for example, which is effective in large groups. But in small societies, where information is valuable and easily lost, teaching works better. Their identification of teaching behavior in an ant shows that a big brain is not necessarily a prerequisite. Thus, surprising as it may seem, Franks et al. were the first to demonstrate the phenomenon of teaching in non-human primates.[5]

It is well known that ants lay pheromone trails to indicate the location of new food sources, but in a remarkable mathematical model they showed that ants might usefully lay *repulsive pheromones*; that is, if ants had thoroughly explored an area and found no food in it, they might deposit a small amount of chemical scent that would discourage their nest-mates

---

5 An individual is a teacher if it modifies its behavior in the presence of a naive observer, at some initial cost to itself, in order to set an example so that the other individual can learn more quickly (Caro and Hauser 1992).

from wasting their time on a fruitless search. However, even though their surprising theoretical result was rigorously based, the team could not find a peer-reviewed journal that would publish it because this sort of behavior is not supposed to happen. They published in a book chapter instead (Stickland et al. 1999). They did not have to wait long before their prediction was experimentally verified. In 2005, negative pheromone marking in ants was demonstrated and published in *Nature* (Robinson et al. 2005).

There are about 10,000 described species of ants. Following the work of Franks et al., however, it would now seem that individual ants are as sophisticated as they need to be, and almost all have high information-processing capabilities at the individual level. It is also possible that ants have collective intelligence too. Back in the early 1990s, the team emphasized collective intelligence above all, and indeed, their very emphasis on determining the simple rules controlling individual behavior indicated that individual ants might be dumb and stupid. This is simply not true. The exciting prospect which Franks and his colleagues intend to pursue and understand, however, is that ants have sophisticated information-handling abilities at all levels from the individual to the collective.

VR 11. Dudley Herschbach, Harvard: "Dimensional Scaling as a New Calculus for Electronic Structure," 1988–1991

I first met Dudley Herschbach at an American Physical Society meeting in New Jersey a few days after the announcement that he had won a share in the Nobel Prize for Chemistry in 1986. I was impressed that he had taken time from his celebrations to come to my talk. I was also wary, because the few Nobelists to contact us seemed to take the view that we should gratefully welcome their participation because it would add luster to our program. But we did not play those fashionable games. We offered freedom, and Nobelists surely had that in abundance. Over the course of the next few months, however, Herschbach proved that he was truly exceptional even by Nobel standards.

Herschbach told me that a few years earlier he had been preparing an introductory course on electronic interactions with nuclei for Harvard chemistry students when he had come across in *Physics Today* an interesting technique used in quantum chromodynamics. He thought it might form the basis of a nice exercise for them, based, as he put it, on making two wrong moves. The first was to calculate the interactions that electrons and nuclei would have if the space they occupied were one-dimensional instead of the usual three. He would then compare that with the interactions expected if the space had an infinite number of dimensions. His reasonable assumption was that the actual interactions in real three-dimensional space should lie between these two extremes, and as the technique seemed to work for the ferociously strong forces between quarks found *inside* nuclei, they should work for the very much weaker ones involving electrons. His back-of-the-envelope calculations made before he asked his students to play his game not only revealed that he was right, but he was astonished to find that his bizarre model (which he called "D-scaling") produced remarkably accurate results. Indeed, they were more accurate than very elaborate conventional calculations using powerful computers! What was Nature trying to tell him? Perhaps she was using Braille?

Conventionally, calculations on the electronic structure of atoms and molecules can be extremely arduous or sometimes even intractable. Accuracy comparable to that provided by the experiment had been achieved only for the simplest systems despite enormous increases in computing power. The fundamental difficulty was the extremely high accuracy required. For the nitrogen molecule, for example, the bond is relatively strong—it is a triple bond involving three pairs of electrons. However, the bond dissociation energy is only 0.04% of the total electronic energy of the molecule. It would therefore be a formidable task to calculate that bond energy to an accuracy of say, 1%. (He compared it to weighing a ship's captain by taking the difference between the ship's weight with him on board and ashore.) As he had quickly found that in some cases he could get significant improvements over the traditional brute-force methods by using D-scaling, he now wished to fully explore what could be done if he took it seriously. Unfortunately, the usual funding agencies had told him, in effect, to stick to his knitting (reaction chemistry) and declined to give him the funding he needed. After an extended series of meetings, largely to test, I have to admit, whether he was really serious about all this, we were delighted to oblige.

As one must be prepared to accept, however, Herschbach's approach might not have "caught on," as he puts it, with the large industry of electronic structure calculators, who now use mostly density functional theory. However, there have been some valuable developments. His former postdoc, Sabre Kais (now at Purdue), has used D-scaling to demonstrate aspects of electronic structure of atoms and molecules akin to phase transitions in statistical mechanics. Marlan Scully (Texas A&M and Princeton) has a sizable group extending D-scaling to treat excited electronic states of atoms and molecules. It would seem that results to a useful accuracy are substantially easier to obtain via D-scaling than by conventional methods.

This somewhat disappointing outcome might be attributable partly to our premature closure. In addition, the BP research director's apparent indifference to Venture Research also meant that even though BP had many chemists with whom Herschbach would have been eager to collaborate (and they with him) opportunities to do so always failed to materialize simply because they needed the research director's approval.

VR 12. Herbert Huppert, Cambridge; and Steve Sparks, Bristol: "Multi-Phase Flows in Dense Media," 1983–1992

Herbert Huppert, a mathematician, and Steve Sparks, a geologist, were both young researchers at Cambridge when they first spoke to us. (We always referred to them as $H^2 + S^2$.) Even at Cambridge, one of the most prestigious research universities in the world, the burden of teaching and administration can be heavy, especially for young people. Their youth and complementary blend of unusual skills had given them an abundance of ideas that they wanted to explore, all of which sounded intriguing and rich with potential. In particular, they had extensive and apparently inexhaustible plans for simple, low-cost experiments that might simulate real planetary processes. Nowadays, as mentioned earlier, one hears constant calls for increased funding because of the rising cost of equipment for world-class research. Freed from the need to compete, however, Huppert and Sparks could concentrate on simple experiments that had never been done. Equipment is rarely on the critical path in such cases. Apart from a modest computer and a video camera, their requirements were for not much more than might have been found in a well-equipped Victorian laboratory, complete with the traditional string and sealing wax. A glass cylinder, for example, containing a cold-water column poised above a hotter one with dissolved chemicals (to make it compositionally more dense) and suitably dyed could probe geologic mixing processes; air forced vertically through a tiny nozzle and containing varying concentrations and diameters of small particles could yield valuable information on erupting volcanoes; and cooling a crystallizing saturated fluid generates convection currents that can lead to complex chemical gradients, and can test how multicomponent systems can solidify and melt. Their enthusiasm was infectious, and radiated "can do" and flair.

Their proposed experiments were designed with elegance, simplicity, and insight. They also created the impression that they could generate new and simple approaches to old problems virtually at will. All they needed was the freedom to get on with it. That involved releasing them from their burdensome university duties while maintaining them in Cambridge's creative environment. The university was very cooperative. All they required was that we provide salaries for replacement lecturers. $H^2 + S^2$ needed very modest funds for equipment and so on, and we were very happy to supply all these things.

Taking full advantage of the flexibility offered by Venture Research funding, and not least the possibility that it could be extended indefinitely, Huppert and Sparks developed some fundamentally new concepts in the earth sciences. Concentrating on dynamical flows in geology, their work addressed such key issues as how rocks melt, how magma evolves chemically and physically in reservoirs under volcanoes, and how magma behaves during volcanic eruptions. The research involved the application of fluid mechanics to geology as well as developing new concepts in fluid mechanics. Thus they played a pioneering role in the creation of the emerging field of *geologic fluid mechanics* (a term they coined). Their work has had major influence on fundamental earth science problems,[6] such as how the crust melts, how magmas chemically differentiate, and how volcanic eruptions proceed. It also had significant economic and commercial implications—for example, their ideas on how, early in Earth's history, very hot lavas melted sediments has led to an understanding on how nickel-ore bodies are formed.

The team influenced not only geology but also fluid dynamics. Starting with honey drooled over toast, for example, surely the most appetizing consumable ever used in an experiment, their work advanced basic understanding of melting and solidification in multicomponent fluid systems, the flow of viscous fluids such as the flow over molten glass during its manufacture, and the propagation of hazardous lava flows following eruption. Their studies of the natural convection caused by phase changes pioneered the idea that composition rather than temperature will dominate behavior in many natural and industrial systems.

Their work attracted widespread recognition in terms of prizes and awards, culminating in their elections to the Royal Society: Huppert became a Fellow in 1987 and Sparks in 1988.

General References
1. H. E. Huppert and R. S. J. Sparks, "The Generation of Granitic Magmas by Intrusion of Basalt into Continental Crust," *Journal of Petrology* 29, no. 3 (1988): 599–624.
2. H. E. Huppert and R. S. J. Sparks, "Cooling and Contamination of Mafic and Ultramafic Magmas During Ascent through Continental Crust," *Earth and Planetary Science Letters* 74, no. 4 (1985): 371–386.

6   The influence of their work has recently been discussed in detail in Davis A. Young's book on the history of igneous geology (Young 2004).

VR 13. Andrew Keller, Ted Atkins, and Peter Barham, Bristol: "Polymer Transition Dynamics," 1984–1990

Polymers are long-chain molecules, but chains are rarely fully extended. Indeed, polymers are normally entangled or closely coiled like a ball of string. When the group first spoke to us, there were no techniques for extending polymers without breaking them. Thus, polymers were difficult to characterize; that is, their strength, stiffness, molecular weight, charge properties, and strength could not easily be measured, which was perhaps *the* major problem in polymer science. Consequently, light or neutron-scattering techniques used to probe polymer structure first had to resolve the confused and tangled large-scale structure before the polymer's molecular properties could be revealed. To say the least, this tended to be something of a nightmare.

Andrew Keller had been a polymer scientist for many years, and had had support from both industry and the research councils. He was also one of the most prolix people I have met, which no doubt created difficulties when he was trying to persuade the various agencies to support his work. Indeed, one of our BP board members referred to Keller's writings as "the Hungarian scrawl." However, I relate this anecdote affectionately, because we found that with patience and perseverance, powerful signals began to emerge from his somewhat circumlocutory ramblings that were worth waiting for. He clearly was very insightful, and was one of the leaders in the polymer field, but it also attracted intense industrial interest, which meant that research—even the basic research—tended to focus on perceived problems. Thus, he had rarely been allowed the freedom to pursue his ideas wherever they might take him without having to constantly justify them in terms of some exploitable outcome. That was precisely the freedom we gave him and his team.

The team proposed to use fluid flows of such high rates that the effects of viscous drag could stretch the chains to their full extent, and thereby enable for the first time their molecular dynamics to be studied.

In 1994 Keller was the winner of Royal Society Rumford Medal. It was awarded "in recognition of his contributions to polymer science, in particular his elucidation of the basis of polymeric crystallization, a fundamental ingredient in many materials, to methods of making strong fibers and to the understanding of polymer solutions which underlie this technology." It was a tribute to his work for Venture Research.

Sadly, Andrew is now deceased.

VR 14. Jeff Kimble, California Institute of Technology: "Quantum Dynamics of Optical Systems," 1983–1992

I first met Jeff Kimble a few years after he had been appointed assistant professor of physics at the University of Texas at Austin. His graduate work had been in Leonard Mandel's laboratory at the University of Rochester, where he had made the first observation of "photon antibunching," a phenomenon that in later years would be recognized as having ushered in a new area in quantum optics. However, the field of quantum optics was then (1977) far from the mainstreams. Kimble therefore had difficulty in securing a postdoctoral appointment, opting instead for a position as a staff scientist at the General Motors Research Laboratories in Warren, Michigan.

Not surprisingly, a management-prescribed research environment was not to Kimble's liking, so he sought employment in academia that offered the freedom to determine his own research directions. But quantum optics was still seen as a backwater, so his wide-ranging search over physics departments in the United States and Canada was fruitless. Fortunately, thanks to Leonard Mandel, he learned of a position at the University of Texas at Austin, and by another stroke of luck a faculty member there (Manfred Fink) remembered him from his undergraduate days at a small college in Texas, whence he became a junior faculty member in 1979.

By the standards of the day, Kimble's basement laboratory, despite his energetic scrounging and lobbying, was not well equipped but was sufficient to launch a modest program. His interest lay in the recently observed phenomenon of optical bistability in which the optical transmission, say, of an optical cavity exhibits well-developed hysteresis effects. The phenomenon was generally well understood from a *semiclassical* perspective, but he wished to exploit it as a means to explore the *quantum* physics of optical systems of "the greatest conceptual simplicity." In this respect his proposed work was against the current trend for device development such as high-speed switches and optical computing. Kimble thought that his difficulties with the NSF stemmed from the fact that his interests fell in the crack between two divisions—the Quantum Electronics Division of the Engineering Directorate, and the Atomic and Molecular Physics Division of the Physics Directorate. The first was not interested in fundamental processes, and there was insufficient atomic physics in the proposal for the second. He had also spoken to the program officers of the Defense Agencies (Office of Naval Research, Air Force Office of Scientific Research, and Army Research Office), who had told him, "If it is not an idea that will make a device work, do not submit a proposal."

The optical system that Kimble wanted to study involved two-level atoms located inside optical cavities. For example, he wanted to investigate the fundamental limits for approaching an instability—that is, the point at which it switches from one stable state to another. In this respect, therefore, he sought to explore the nature and impact of quantum tunneling for bistable systems. Ultimately, his goal was to move optical physics into a new regime where manifestly quantum processes become dominant. His strategy was to develop ways of exploring the wide gulf between the macroscopic world, for which our normal intuition suffices, and the world of quantum mechanics, which seems to be governed by a bizarre set of non-intuitive rules. As we talked in his lab, he said that if the tabletop next to us represented the full domain of quantum mechanics, then in his view our current understanding was limited to only one tiny patch in one corner, with the rest unexplored and ripe for discovery.

We talked about other problems in the field of quantum mechanics such as how to measure a system without disturbing it, and the generation of new types of quantum states that have less fluctuations (noise) than does the vacuum state. In other words, they are "darker than darkness," as he imaginatively described them; states that offered the prospect of making measurements more accurately than the Heisenberg uncertainty principle allowed in its usual (naive) application. Here was a researcher who seemed equally at home with either theory or experiment. He also convinced us that given freedom he intended to forage deep into what seemed to be an exciting and unexplored wilderness, which is precisely the sort of mission we love to support. We were happy to oblige, therefore, and give him his first major grant.

Kimble's early experiments in optical bistability laid the principal foundations for the modern field of cavity quantum electrodynamics (QED). His Venture Research support led to the first observations of the interactions between single atoms and photons. These included the demonstration of a quantum phase gate in cavity QED for quantum logic with single photons (in 1995) and the realization of a laser that operates with one, solitary atom in a regime of strong coupling (in 2003). Here, Kimble has pushed laser operation to its conceptual limit.

In 1986, with his second Venture Research grant (and later continuing with his third), Kimble and his colleagues embarked on a new direction, namely, the creation of "squeezed light" with quantum fluctuations below the standard quantum limit. Using the freedom we allowed and indeed encouraged, Kimble courageously spent his entire three-year budget in a few months to implement his idea for the creation of squeezed light via parametric down-conversion. The result was a tour de force, and

Citations in Each Year

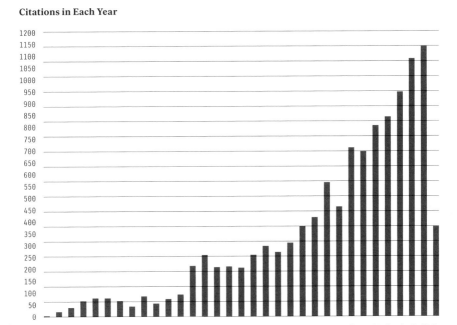

Figure 14: H. J. Kimble's citations in the literature. (Source: ISI Web of Science. Copyright 2007, The Thompson Corporation.)

provided what is still the most successful route for the generation of squeezed light more than 20 years later. Moreover, Kimble discovered a direct demonstration of the Heisenberg uncertainty principle for light, and made the first measurements with sensitivities beyond the standard quantum limits. His work with squeezed light led to the first realization of the Einstein–Podolsky–Rosen paradox (dating to 1935), and to the first demonstration of bona fide quantum teleportation (in 1998).

Kimble was elected to membership in the US National Academy of Sciences in 2002.[7]

---

7   Ten years after his Venture Research support ended, Kimble sent me a personal note on the occasion of his election. He said:

I hope that you will remember many years ago when you came to UT to speak about the VRU, and following that, the receipt of a letter from a young assistant professor at UT. One thing led to another, and eventually to my VRU grant, which has made all the difference in my career. Back then the things that I proposed were well outside the mainstream of traditional AMO physics, but by now have become established (credible) programs within a broader scientific community. Certainly, not everyone was supportive of me or my ideas. So, all these many years later, I do sincerely thank you for taking a chance on me, and for making what was then, a decision far outside the fashion of the day. It has made all the difference for me.

Kimble's experience provides further demonstration that the careers of putative members of a twenty-first-century Planck Club cannot be expected to run according to bureaucratic or indeed any externally set agenda. His citations rocketed (see **Figure 14**) following his Venture Research support even though his interests were in a backwater when he started out.

VR 15. Graham Parkhouse, Parkhouse Consultants: "An Integrated Approach to the Theory of Structures," 1986–1992

Is it possible to do Venture Research work in engineering? Research in the natural sciences should always imply the possibility of a surprising outcome, however mundane it may be initially. The question therefore arises: Is it possible to conceive of open-ended research in engineering since highly specific objectives seem essential? I thought it was, but our problem was that we got very few proposals from engineers.

In an attempt to encourage them, I published an essay in the *New Scientist* on blue skies research in engineering and the role of the *theoretical engineer* (Braben 1986). Imagine that you ask an engineer to design a bridge or a building. The engineer will consider the specification, visit the site, think about costs and timescales, and generally contemplate the problem. Then at some fateful moment, the engineer will pick up a pencil and tentatively sketch a design and will then probably have a clear mental picture of the methods of construction, location, materials to be used, and so on, and the simple act of sketching them might have closed the other options. Henceforth, the engineer draws either on well-established design processes, or others that may have to be derived if the problems are substantial—the Sydney Opera House,[8] for example—but the objectives will be clear. However, when we come to consider what might have happened during the interval between the initial invitation and the engineer starting to sketch, we can say very little. Do these *abstract* processes (such as those that Parkhouse illustrated in **Figure 15**) have structure? How might they vary between recently qualified engineers and those with, say, 20 years' experience? My suggestion was that those who would address such questions should be called "theoretical engineers" as their research might lead to new types of things being done, and perhaps to surprising outcomes.

Parkhouse was a consulting engineer in a large company when he first came to see us. Not surprisingly, he lacked the freedom to develop his

---

8   Initially designed in 1956 by the 38-year-old Danish architect Jørn Utzon, the building of this iconic structure by the British company Ove Arup, which was completed only in 1973, required many technological innovations that have since passed into accepted engineering practice.

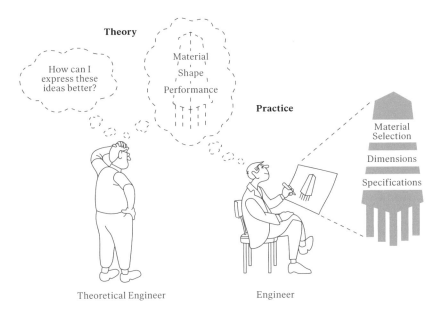

Figure 15: The role of a *theoretical* engineer. While engineers use existing theory to produce real structures, the theoretical engineer is considering how to improve the conceptual structures that the theory implies.

theoretical ideas on the nature of engineering processes. He explained that he was interested in the concept of hierarchy in design and wanted to build experimental rigs that would test his ideas with real materials. However, although we were soon convinced that he was a Venture Researcher, it was not possible for him to participate in our scheme while remaining with his company. Showing remarkable courage and conviction, therefore, he decided to resign if we would support him, which we were delighted to do. Indeed, he turned out to be the only industrial scientist we could find. But first we had to find him a locus, and luckily, we were able to persuade the University of Surrey (the university nearest to his home) to offer him an academic post and a lab. He could now begin his life as a "theoretical engineer," and develop the integrated approach to the theory of structures that he had long contemplated (Parkhouse and Sepangi 1994).

Parkhouse's thinking starts with **Figure 16**. It shows unrecognized relationships between shape and material, but he gives these concepts special meaning. First, they are both abstract concepts in his theory (Parkhouse 1987). *Shape* is a purely geometric concept without substance, while *material* is unbounded shapeless substance. The mathematics he associates with shape is simply geometry. The properties he associates with

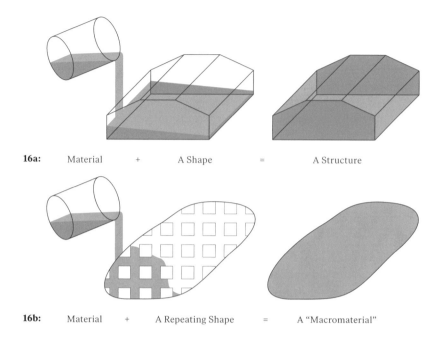

16a: Material   +   A Shape   =   A Structure

16b: Material   +   A Repeating Shape   =   A "Macromaterial"

Figure 16: Two important relationships between material and shape.

---

material include density, stiffness, strength, and cost. In everyday engineering they always coexist, as shown in **Figure 16a**. Every structure can
be characterized by shape, material, and geometry. Such properties as the
strength and cost of a bridge, for example, describe a structure's performance, and the designer's task is to choose a structure whose properties
meet the specification. Choosing a material—steel, aluminum, concrete,
GRP, or timber, say—is relatively simple, but then one is faced with the difficult problem of choosing the shape as there are an infinite number of
possible types: lattice, tubular sections, stiffened plates, and so on. Almost
all the geometric complexity comes from the repeating shapes, but when
filled with a material, as in **Figure 16b**, they have material properties rather than structural ones. In effect, they become another material, which
Parkhouse calls a *macromaterial*, which is less dense than the original because of the space entrained within it. Thus, he shows the synthesized
material in **Figure 16** in a lighter shade of red, and describes the structuring (the shaping of material into tubes, lattices, etc.) as a process of *material dilution* (Parkhouse 1984). Solid beams of macromaterials of the
correct strength and density, for example, can represent rolled-steel
joists—timber would be a good choice.

Macromaterials are hierarchical. The repeating shapes in **Figure 16b**, for example, can exist in more complex or lighter forms, each with a specific density, stiffness, strength and cost. If designers had this kind of information (coming from, say, an international program of testing and analysis on a range of macromaterials), their tasks could be transformed into a rational process of macromaterial selection guided by performance and cost rather than by pragmatism.

His Venture Research work focused on the extraction of data on macromaterial strength and stiffness from load tests on structures. As it progressed, it became clear that some types of repeating shape are suitable for use in compression, while others are suitable for tension. A fibrous form, for example, is excellent in tension and poor in compression, while a block or granular form is excellent in compression and poor in tension. Materials likewise fall into two categories: those that are ductile and excellent in tension, and those that are brittle and excellent in compression. The performance of a prestressed composite member of ceramic blocks held in compression by taut GRP wrapping was shown to exhibit ductile behavior in tension even though neither of the constituent materials was ductile (Parkhouse 1989). He discovered a remarkable lenticular shape that had the geometric attributes to qualify it as a model structural member (Parkhouse et al. 1992). Under any endloading this model member *has one identical stress pattern at every section along its length* even when it is approaching failure and the material within it is yielding. Such a serendipitous shape would be a gift to theoretical engineers wanting to understand the structural implications of material nonlinearity.

Sadly, Parkhouse's transformative research came to an end following our closure. Little support was available elsewhere as engineering is understandably a conservative field and there had simply been insufficient time for his radical ideas to take root. That is a great pity. In the short term, they could be used as an aid for teaching as they enable a perspective on every type of structure that has ever been built. In the long term, his ideas would offer the prospect of applying the experience of building in steel and concrete, say, to all modern high-strength materials. Above all, however, he has made the remarkable discovery that the performance of a structure can be assessed by enveloping it with an imaginary shape, and treating its contents as a form of "material" independently of all its countless components, a technique that could be applied to all engineering fields and to technology in general.

VR 16. Alan Paton, Eunice Allan, and Anne Glover, Aberdeen: "Induction of Novel Symbioses between Bacteria and Higher Organisms," 1982–1992

When Alan Paton first came to us in 1982 he was already semiretired (he was born in 1922). About 10 years earlier, he had examined rotten potato cells under a microscope. He had found many dead cells, of course, but some were still living even though they were almost full-to-bursting with bacteria. He asked himself the simple question: How can these bacteria get into cells when they cannot get out? He then recalled a private communication he had received years before about L-form bacteria—bacteria that have either temporarily or permanently lost the ability to synthesize their cell walls.[9] The walls not only determine a cell's shape but also protect the cell from attack. Deprived of it, they can take up virtually any shape, and Paton thought that the L-form's plasticity may have enabled the bacteria to invade the potato cells. However, that only deepened the mystery. The bacteria in question were pathogens, but cells apparently full of them still seemed viable. What was happening? Had he discovered a new type of organism that was neither wholly plant nor wholly bacterium? Had he come across a new plant-bacterium symbiosis?

He had tried unsuccessfully to get funding, but the "theoretical dogma hitherto universally accepted as correct" was that pathogens and plants do not enter into symbiotic associations. The dogma forbade it. The major problem was that plants are eukaryotes while bacteria are prokaryotes. They use different genetic languages, therefore, and so the dogma maintained that they cannot easily communicate. Paton was also a self-styled, "good old-fashioned microscopist," and microscopy was no longer fashionable. His peers' response implied that his eyes might be deceiving him so that what he thought he was seeing was really an artifact. That would normally have been the end of the matter, but he was undaunted and struggled on intermittently with whatever meager resources he could scrounge. These later studies reinforced his view that he was seeing an unusual symbiosis. Furthermore, they seemed to show that plants "inoculated" with pathogenic L-forms were resistant to infection from fully active bacteria.

We liked their proposal, but the weight of opposition was such that we were initially only able to offer short-term support (i.e., one year). Part of the problem was that he could not always reproduce some of his results.

---

9   These bacterial forms were first described at the Lister Institute of Preventive Medicine (a school of the University of London) in 1935.

He was as mystified by this as we were. However, his vast experience convinced him that something very unusual was going on, and he was determined to understand it. We made several visits to Aberdeen, where, with Paton's microscope, we could actually see pathogenic bacteria apparently thriving within cells that should be dead. There seemed little doubt, therefore, that something very interesting was going on. We knew that normal cells contain organelles such as mitochondria and chloroplasts that seem bacterial in origin. Despite current dogma, it was possible that Paton might be about to rediscover a form of association that Nature had experimented with eons ago, but that she no longer needed. The lack of reproducibility might stem from his failure to identify every contributing factor, which is often a problem in the biological sciences. Thus, unknown processes were influencing the phenomena he was observing, and as they would therefore be uncontrolled, they might be the cause of the troublesome lack of reproducibility.

All this was indeed intriguing, and we decided to go ahead with full-scale support subject to the proviso that the team should recruit a molecular biologist to help increase the breadth of their attack, and perhaps also silence some of the skeptics by putting his results on a more acceptably rigorous footing. Thus, Anne Glover subsequently joined the group with expertise that added a powerful new string to their bow.

The major problem was to demonstrate beyond doubt that pathogenic bacteria were fully metabolizing within the plants, that is, that *both* organisms were thriving. This was done by means of one of the most elegant experiments performed during the 10 years of Venture Research. The key part of the experiment was their use of the *lux* genes—the name given to genes expressing the enzyme luciferase. The *lux* genes can be isolated, and using genetic engineering techniques can be introduced into any bacterial cell. It emits tiny flashes of light if the *lux* genes are expressed within a normally functioning cell. Their experiments are illustrated by the cartoons in **Figure 17**.

Thus, the L-form of the pathogenic bacterium seems to have lost its ability to infect (Amijee et al. 1992; Walker et al. 2002), and the inoculated plant can grow and develop normally. Unfortunately, these experiments were completed only in 1990, shortly before Venture Research's demise. Other experiments undertaken by Eunice Allan et al. using a different reporter mechanism than *lux* (the *gus* β-glucuronidase reporter gene) revealed that plants inoculated with L-form bacteria derived from nonpathogens also conferred resistance to both bacterial and fungal plant pathogens. As the *gus* gene is under the control of a bacterial promoter, its expression can take place only in a prokaryotic environment. But plants

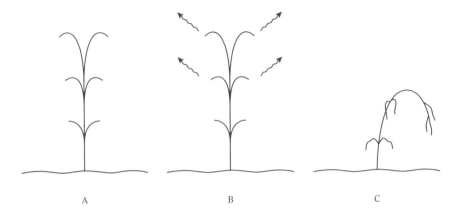

Figure 17: Cartoons illustrating the effect of inoculating a plant with either the L-form or the normal form of a pathogenic bacterium: (a) French bean, *Phaseolus vulgaris*, two days after being inoculated with the L-form of the pathogen *Pseudomonas syringae*; (b) French bean two days after being inoculated with L-form of the pathogen equipped with the *lux* genes; (c) French bean two days after being inoculated with the normal pathogen equipped with the *lux* genes.

are eukaryotes, of course. Their discoveries would therefore seem to have established a previously unexpected form of symbiosis, and could open up a wide range of scientific possibilities. However, although their discoveries are potentially transformative, they have yet to be widely recognized. As funding has been highly problematic, the team has split up. Eunice Allan continues with the quest on very limited resources, but little progress with the basic research has been made since the Venture Research funding ended.

Sadly, Alan is now deceased.

General References
1. E. Tsomlexoglou, P. W. Daulagala, G. W. Gooday, et al., "Molecular Detection and β-Glucuronidase Expression of *Gus*-Marked *Bacillus Subtilis* L-Form Bacteria in Developing Chinese Cabbage Seedlings," *Journal of Applied Microbiology* 95, no. 2 (2003): 218–224.
2. P. W. Daulagala and E. J. Allan, "L-Form Bacteria of Pseudomonas Syringae pv Phaseolicola Induce Chitinases and Enhance Resistance to Botrytis Cinerea Infection in Chinese Cabbage," *Physiological and Molecular Plant Biology* 62, no. 5 (2003): 253–263.

VR 17. John Pendry, Imperial College of Science and Technology: "Transport in Disordered Systems," 1982–1991

I first met Pendry shortly after his appointment as professor of theoretical solid-state physics at Imperial College at the age of 38. Pendry was very much a rising star, and as such well funded by the Science Research

Council, as it then was. However, he quite naturally saw his appointment as presenting an opportunity to branch out into new and unexplored areas as so many scientists before him had done on these occasions. Change can be a powerful catalyst and source of inspiration. But since 1970 or so, funding in the United Kingdom has been managed nationally, with only modest funds available for individual universities to disburse as they see fit. Astonishingly, therefore, support that would allow Pendry to change tack and to vary it as his research developed was highly unlikely to be forthcoming.

The electronic properties of crystalline materials were well understood at that time. However, it was normally assumed that local-scale behavior—involving a few tens of molecules, say—would be invariable throughout the crystal. In other words, the crystalline materials must be virtually perfect. This meant that the theory had little practical value when applied to real materials as they are usually far from perfect. Electrical conductors have electrical resistance. That resistance, however, is not an intrinsic property of the material, but is a function of the imperfections, thermal fluctuations, and disorders unavoidably introduced during manufacture. Amorphous materials were also becoming increasingly important—for use in solar cells, for example—and the influence of disorder on their conducting properties presented some of the deepest problems in theoretical physics. The understanding of liquids was also in its infancy.

Thus, theoretical understanding of electronic properties was almost entirely confined to ideal materials—that is, there was no general theory describing the interaction of waves (e.g., the flow of electrons) with disorder. Pendry wanted to exploit the freedom that our support offered to embark on an ambitious study aimed at understanding the electronic properties of real materials, and to combine his theoretical studies with the large-scale use of computers to simulate their properties. Hopefully, it would lead to an understanding of the disordered state, and thereby help lay the foundations for the next generation of electronic devices.

Venture Research support indeed proved indispensable in enabling Pendry to found a leading new condensed-matter theory group. He applied group-theoretic methods to transport in disordered systems to give a complete solution of the general scattering problem in one dimension, and derived advanced techniques for treating higher dimensions (MacKinnon et al. 1992; Pendry 1994). A key result was the prediction that in all dimensions the channels for transport show a bimodal distribution—they are either open or closed in the limit of large systems.

John Pendry was elected a Fellow of the Royal Society in 1984. His services to science were recognized with a knighthood in 2002.

VR 18. John C. Polanyi, Toronto: "Surface Aligned Photochemistry," 1983–1991

When I first met John Polanyi in 1983, he was already one of Canada's most distinguished scientists. His long-term but once-pioneering work on reaction dynamics was reasonably well funded, but Polanyi now wished to turn his attention to the exploration of a new area—the possibility of catalyzing photochemical processes by the use of carefully chosen surfaces. Sadly, the common approach to exploratory research in academia nowadays is for a senior scientist to ask a postdoc to look at it for a few years to see what might be done with more significant resources. One can keep lots of irons in the fire in this way, which in the present funding regime is usually vital. The drawback is, of course, that the postdoc charged with the reconnaissance is not only inexperienced but usually looking to his or her next job. Subtle signs may easily be missed. None of this applied to Polanyi. He was convinced that the new area was going to be important, and wanted to devote most, if not all, of his energies to it. However, this "embryo field," as he described it, stood little chance of being funded by the research councils because the consensus view was that photoprocesses at surfaces would be inefficient and complicated by energy transfers to the surface. Polanyi did not agree. His theoretical studies had indicated that his approach—that is, surface-aligned photochemistry—could be viable.

Catalysis is one of the most important fields in chemistry, but in the absence of complete understanding, it is dominated by empirical approaches. Polanyi was not so much interested in catalysis per se but in the fundamental processes that go on at surfaces. In particular, he was interested in their molecular dynamics, and the patterns of molecular motions that convert reagents into products. This would involve studying reagents and products as a function of energy and direction (with respect to the surface) of the photorecoiling molecules adsorbed on the surface. As a one-time elementary-particle physicist, I knew that this was a powerful technique. It would not be easy to make it work in the more complex chemical domain, but if anyone could, it would be Polanyi. We were very happy, therefore, to fund his change of tack.

His initial idea was that he could both simplify and reveal the dynamics of having the reacting molecules "pinned" by weak physisorption forces to a surface before shining light on them to initiate a reaction. Thus, the reacting molecules at the surface would be aligned and separated by known distances, and by varying these quantities they could "catalyze" the reaction in (ultimately) a predictable way (Bourdon et al. 1984). The results showed that an absorbed bromide molecule sent a shower of bromine atoms upward at a defined angle to the surface, aligned along the axis of the molecule originally placed on the surface. In the course of the next 30 papers, they learned how to aim the molecular shower along the surface rather than away from it, and hence achieve their "surface-aligned" objective.

In 1981, G. Binnig and H. Rohrer invented the scanning tunneling microscope (for which they won a share in the Nobel Prize for Physics in 1986). This led Polanyi's group to realize that they could use this technique (which revealed individual surface atoms) to study the dynamics of photoinduced surface reactions one molecule at a time. The new technique also provided a novel means of imprinting desired patterns of molecular dimensions at surfaces.

Today (2007) the group allow molecules to self-assemble into patterns on silicon surfaces, and then permanently imprint those patterns through the agency of light. It is as if they could photocopy molecular systems only a few molecules wide (Dobrin et al. 2007). The group's new direction started with Venture Research, but now seems about to benefit the makers of nanostructures and, perhaps, nanocatalysts.

Polanyi won a share in the Nobel Prize for Chemistry in 1986 for his pioneering work on reaction dynamics.

VR 19. Martyn Poliakoff, Nottingham: "Supercritical Fluids: An Environment for Reaction Chemistry," 1988–1991

As I have stressed, promoters of Venture Research need to be constantly on the alert for intrepid explorers yearning for opportunities that would allow them to penetrate deeply into the unknown, or to bring about the downfall of a cherished theory in an important area. But slavish adherence to any strategy is normally a bad thing in research, and indeed, one should always be on the lookout for *anything* unusual, even if at first glance it may seem mundane. That includes, for example, ideas that might seem obvious, and on which one might comment: Surely that's been done before?

Ordinary liquids are virtually incompressible, of course, while gases have low densities. Hitherto, despite these serious disadvantages, chemistry had usually been performed in gaseous or liquid environments. Poliakoff was planning a major departure from this historic practice by proposing a possibly revolutionary step of breathtaking simplicity: that of examining the potential of supercritical fluids (see **Poster 20**) as *an environment* in which to do reaction chemistry (see also **Figure 18** and **Table 14**). Previously, users of supercritical fluids had confined themselves to exploiting their specific properties. Industry had used them for decades; the oldest example was the use of liquid CO_2 for extracting caffeine from coffee. However, there was a good reason why the potential of the supercritical regime had been largely unexplored—it was necessary to confine the fluid in a high-pressure vessel, while at the same time retaining the ability to observe and study its behavior. While these obstacles are formidable, they are not insurmountable, but they do require considerable expertise. The general view had always been that there seemed to be no obvious prizes in terms of new products or processes that would justify the effort. Why bother, therefore?

Poster 20

—

Poliakoff and Supercritical Fluids

A generalized phase diagram for a pure material is shown in **Figure 18**. The materials that Poliakoff proposed to use—ethane, carbon dioxide, or xenon—are usually known as **gases**. If the temperature and pressure are high enough, however, the average distance between molecules is reduced to such an extent that the material—by now a supercritical fluid—acquires some of the properties associated with liquids. At the same time, molecular kinetic energy is still high enough for the material to retain some gas-like properties.

As **Figure 18** shows, a material can become supercritical only if the temperatures and pressures to which it is subject are above certain minimum values, T_C and P_C, respectively. Supercritical fluids are significantly denser than gases (see **Table 14**) but usually less dense than liquids. Density is a key factor in the ability of a fluid to dissolve solid materials. In sharp contrast to liquids, supercritical fluids are highly compressible; a relatively small change in applied pressure can cause an appreciable change in the density of the fluid. Thus, the solubility of a given material in a supercritical fluid becomes a function of the applied pressure.

Poliakoff thought differently and had acquired the necessary expertise. He argued that the unprecedented degrees of control allowed in the supercritical regime could open up new types of chemistry, as well as leading to new insights on the old. For example, the physical properties of supercritical fluids are such that spectroscopic studies throughout the range from ultraviolet to infrared become possible, and nuclear magnetic

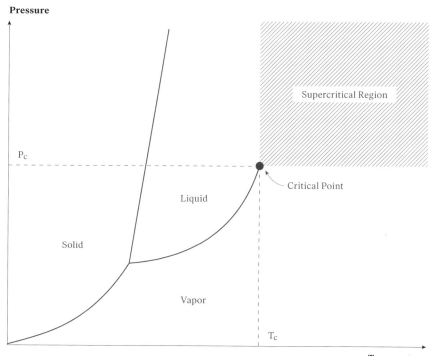

Figure 18: A Generalized Phase Diagram for a Pure Material.

Table 14: Some Fluids Having Critical Points Near Room Temperature

Material	Temperature T_c, °C	Pressure P_c, atm	Critical Density, g/mL
Ethylene	9	50	0.23
Xenon	17	59	1.1
Carbon dioxide	31	74	0.47
Ethane	32	49	0.20
Water	374	218	0.32

resonance spectra should become sharper and easier to interpret. Unfortunately, this was not enough for the research councils, but they had not rejected him completely. The Science Research Council had provided modest funds from an industrial-collaboration scheme (with ICI), but it offered neither the freedom nor the funds adequately to support his revolutionary steps. Supercritical fluids had fascinated scientists for more than a century, but Poliakoff's vision, enthusiasm, and novel ideas on their use strongly appealed to us.

With funding no longer a limitation, Poliakoff's first task was to develop equipment to handle high-pressure fluids safely in the lab. He and his team devised an ingenious modular system of magnetically activated taps, cells, and gauges that could be quickly built up into the apparatus needed for a particular experiment (Poliakoff et al. 1995). Chemically, he pioneered a completely new field—organometallic chemistry in supercritical fluids—that has now been taken up worldwide, particularly for homogeneous catalysis in supercritical CO_2.

Much of Poliakoff's work exploited the fact that supercritical fluids are gases, and so they are completely miscible with permanent gases such as hydrogen or nitrogen. This enabled him and his two Venture Research fellows, Steve Howdle and Mike George, to make a series of new metal dihydrogen and dinitrogen compounds (Howdle et al. 1990), the unexpected stability of which provided a challenge to theoreticians. Indeed, Poliakoff then demonstrated that these compounds could be generated and isolated using a small-scale continuous reactor, rather like a tiny chemical plant (Banister et al. 1995; Darr and Poliakoff 1999).

The enforced end of Poliakoff's Venture Research funding coincided with the birth of green chemistry in the United States. "Green chemistry" is a new field in which researchers aim to invent radically cleaner approaches to the design, manufacture, and use of chemicals (Poliakoff et al. 2002). Thus, Venture Research put Poliakoff in a position where he could exploit the growing interest in this field and in "clean technology," as Ken Seddon (VR 22) was also doing. In 1994, Poliakoff was awarded the first Science and Engineering Research Council's Clean Technology Fellowship, and Seddon won the same award a year later. In 1995, Poliakoff started a major industrial collaboration that culminated in the building of a full-scale, 1,000-ton/year chemical plant, the world's first multipurpose supercritical chemical reactor (Licence et al. 2003), a direct spin-off of his work with Venture Research. A second collaboration involved selective partial oxidation in supercritical water, where the key factor is the miscibility of oxygen with the water and reactants. This work (Fraga-Dubreuil and Poliakoff 2006) is leading to a new approach to manufacturing terephthalic acid, one of the chemical precursors to the now-ubiquitous plastic water bottle, which might displace the current oxidation technology that has dominated for more than 50 years. These and other industrial collaborations have placed Poliakoff among the world leaders in green chemistry.

Poliakoff was elected a Fellow of the Royal Society in 2004 with the citation specifically mentioning his achievements in green chemistry.

Venture Research completely changed Poliakoff's research direction, which is now devoted entirely to supercritical fluids. It launched his two young colleagues, Steve Howdle and Mike George, on their scientific careers. Both are now full professors and still work with these fluids. It made the United Kingdom a major center for supercritical research, and many of Poliakoff's students and postdocs work in academia and industry, both in the United Kingdom and internationally. More generally Poliakoff says that his Venture Research has been a big factor in creating the current level of interest worldwide in supercritical fluids as reaction media.

VR 20. Alan Rayner and John Beeching, Bath: "Towards Understanding Multicellularity," 1987–1990

At our first meeting, Alan Rayner described the vast unexplored terrain dominated by the fungi. Much previous genetic and biochemical work had been on the commercially valuable unicellular fungus, *Saccharomyces cerevisiae* (brewer's yeast), but such unicellular forms are greatly outnumbered in nature by filamentous forms. These latter fungi produce *mycelia*, labyrinths of protoplasm-filled tubes, which, although perhaps resembling amorphous masses of cotton wool, are actually very complex. They occupy a wide range of habitats, are more species-diverse than plants are, and probably outweigh the animal kingdom. Forests, grasslands, and heath could not exist without their symbiotic support, and some forest forms are among the largest living organisms found on Earth. To thrive, these fungi need to be able to forage to find new food sources, to move to them when the old resources are depleted, and to hold onto and expand their genetic territory in the presence of other organisms.[10] Correspondingly, they exhibit remarkable developmental versatility, display senescence, and are able to switch between vegetative and reproductive growth.[11]

While fungi are interesting in their own right, they are also among the simplest forms of eukaryotes. Nature tends to be conservative, so tricks used in one domain tend to be repeated in similar domains. The team had recognized, therefore, that the fungi would be a highly versatile and convenient model eukaryotic system amenable to experimentation. Using a combination of whole organism and molecular biological approaches, the team wanted to study gene expression and regulation; how mycelia divide their labor, coordinate their development, and interact with self and nonself; and a huge range of biochemical pathways that would be relevant to eukaryotic organisms in general—animals and plants.

---

10 The familiar mushrooms are the reproductive structures of particular mycelial species.
11 Fungi produce chemicals that enable them to retain their own and enter others' territory, and can reject or accept potential mating partners.

However, the study of mycelia from these generic perspectives had been almost totally neglected.

The emergence of multicellularity is one of the major landmarks in evolution. It introduced a virtually boundless scope for the development and diversification of life, a rigorous understanding of which would be one of the most important goals in biology. However, the general principles underlying the ways in which eukaryotic organisms coordinate the development and interrelationships of distinctive cell and tissue types, which are at the heart of multicellularity, were not understood. The scope of their proposed studies was perhaps greater than could easily be managed, even for a determined group. But ambition should not be penalized, and given freedom, they would soon home in on the topics that most interested them. I could hardly believe that the research councils would not fund such a fascinating study, and we were delighted to do so.

Using an elegant technique—the examination of interactions between mycelia of the same species but obtained from different parts of the world—the team quickly made a variety of discoveries with implications for understanding how cells and their organelles respond to one another and to their environments. Their molecular biology showed that these interactions could have highly unstable outcomes, resulting in genetic takeover, degeneration, and chaotic patterns of development. Especially spectacular was an interaction between an English strain and an Australian strain of the same fungus, which led to the cellular degeneration of the latter accompanied by the outgrowth of a mass of crystalline aggregates of a single, *optically pure*, organic compound: (+)-torreyol. These organic crystals had remarkable fluid-dynamic properties, growing up to six centimeters long and branching in a wide variety of patterns reminiscent of fungal mycelia themselves.

As I have mentioned, one of the advantages of Venture Research was the frequent conferences and workshops at which researchers could share their heretical ideas. Rayner and Beeching met Ian Ross (see VR 21) at one of these meetings, and quickly realized that many aspects of their ideas complemented each other, and indeed, although their routes were different, they had common objectives. They both used fungi as model systems, and had noticed that mating in some fungi resulted in the formation of two functional zygotic thalli, each with identical nuclei but with different mitochondria and phenotypes. Thus, they discovered the potential role of mitochondria in regulating cell behavior a decade before the current paradigm of mitochondrial control of programmed cell death (apoptosis) and other senescence programs were accepted (Olson and Stenlid 2001). They subsequently expanded their hypothesis to include a mitochondrial role in regulating specific nuclear gene expression, which if perturbed could lead to cell and whole-organ senescence and possibly explain the apparent random nature of causes of death, even among close siblings and clones.

It was some years before Rayner and Beeching—still keeping in touch with Ross—were able to follow up on these ideas following Venture Research's closure. Eventually, together with Zac Watkins, they obtained modest funding that enabled them to gather data to support their hypothesis that fungal mycelia are complex nonlinear systems whose variable patterns of cellular development and response arise from distinctive expressions of oxidative and antioxidative metabolism. These distinctive expressions can both induce and suppress cellular degeneration, and lead to variations in the permeability, deformability, and continuity of cell boundaries.

Ultimately, Venture Research enabled Rayner and Beeching to develop a systemic model for understanding the development and ecology of fungi in terms of the generation of variable boundary chemistry in dynamic relation with the oxidative and reductive potential of their habitat. Unfortunately, they have found it difficult "in a scientific community obsessed with specific mechanism and detail to the point of blindness to underlying systemic process," as they put it, to gain either recognition of or further support for their work. They were obliged to abandon their mycological research in 1999, a year after Rayner had been president of the British Mycological Society.

Beeching subsequently pioneered an investigation into the molecular genetics and developmental biology of the rapid postharvest deterioration of cassava roots, one of the most important tropical food crops. Hitherto, cassava had been regarded as too uneconomic in first-world terms, and the problem too complex to be worth bothering with. Now others have added their weight to Beeching's initiative. The phenomenon of variable apoptotic responses to oxidative stress, which the group pioneered in their Venture Research days, is now at the heart of the current global strategy on cassava development.

VR 21. Ian Ross, University of California at Santa Barbara: "Cytoplasmic Control of Nuclear Behaviour," 1988–1991

My first contact with Ian Ross stemmed from an act of desperation, as he put it, when, with only the cost of a postage stamp to lose, he sent a tentative letter of inquiry to our London office. His interests were in developmental biology, but he saw enormous advantages in using fungi as a model system. As these two areas of science were generally thought to have little or no overlap, the funding agencies had always declined his proposals perhaps for the simple reason that there were few peers to review them.

Ross explained that the development of a fertilized egg (a single cell) into a multicellular adult is the result of a well-choreographed program of genetic, biochemical, and cytological events that when initiated causes them to operate in the correct temporal sequence. In eukaryotic cells, it is normally presumed that the cytoplasm is controlled by the nucleus. However, these roles are reversed in the first stages of embryo development. In the earliest stages, nuclei are totipotent and capable of expressing any part of their genome. (Today, this quality of totipotency is at the core of the highly fashionable field of stem cell research.) A short time later, these same nuclei in animals and plants become highly specialized and express only the particular genes appropriate for the next sequence of events in that part of the embryo—liver cells, for example, in humans. The trigger for the change in behavior comes from spatially localized molecules in the cytoplasm rather than the nuclei. However, the study of this essential nuclear programming is difficult in animals and plants because the changes take place rapidly and the cells are very small. On the other hand, cells at the tips of filamentous fungi offer the enormous advantage that *the embryonic stage persists virtually indefinitely*, and the cells can be large. He believed, therefore, that the filamentous fungi offer a radically new way of studying these early stages of development, and would open up a new set of investigations into the cytoplasmic control of nuclear behavior.

Working on shoestring budgets, he had already shown that the totipotent nuclei of fungi (in fungi, individual cells may have many nuclei) enter specific developmental pathways only if they are in areas of cytoplasm that contain the appropriate macromolecules; that is, these nuclei are regulated in the same way as the nuclei of early embryos. Not only had he confirmed, therefore, that fungi were a viable model system; it was also one that is readily amenable to experimentation. Regulation had to date only been studied in far less accessible systems.

He argued that his proposed study would permit a direct means of controlling the constituents of the cell membrane, and could thereby lead to new ways of regulating pre- and postembryonic differentiation of all higher plants and animals, and of controlling many plant diseases. The mechanisms he expected to discover might also be useful in regulating and reactivating cells of adult organisms normally unable to regenerate (e.g., nerve tissue). His substantial list of objectives may or may not have been achievable, of course, but the clinching argument for us in Venture Research was that Ross was not really interested in any particular problem but wanted to understand some of the complexities of developmental biology. He was merely trying to make his proposed research look as attractive as possible, as if he were applying to a conventional agency. From our point of view, however, he had a most imaginative and new approach for tackling one of the major unsolved problems in biology, and we were therefore delighted to support it.

Ross's collaborative work with another Venture Research group is discussed at VR 20.

Ross continued his work with intermittent small-scale funding after 1991. In September 2003, he presented a paper to the 10th Congress of the International Association of Biomedical Gerontology on the evolutionary reasons for the fact that mitochondria are inherited only from one parent—the mother (Ross 2004). As he explains:

> Because mating brings two sets of haploid nuclear genes into the zygote, two alleles of each of the proteins in question could theoretically be expressed. If maternal and paternal alleles differed in base sequence affecting amino acid sequence, the two allelic proteins in zygotes could have different abilities to interact with the mitochondrial-gene-coded proteins of the complexes. Competition, therefore, could lead to adverse interactions and less than optimal mitochondrial efficiency.

Mitochondria play dynamic roles, affecting programmed cell death, numerous diseases, and possibly cell aging and senescence leading to Alzheimer's disease and cancer. Currently, the focus has been on the nuclear control of mitochondria, but the reverse is also possible. His hypothesis was that the onset of cell aging may be a result of a failure of the mitochondrial influence on nuclear genes.[12] Unfortunately, he has been unable to obtain further funding for this line of inquiry until its validity has been established, yet another example of the catch-22 nature of modern funding.

---

12 This idea was discussed by Nick Lane (2005), who added that this kind of gene regulation of selected alleles might be a kind of imprinting.

VR 22. Ken Seddon, The Queen's University of Belfast: "Chemistry and Physics in Ionic Liquids," 1988–1991

Our first meeting with Ken Seddon as a principal investigator was in our plush, City of London headquarters.[13] He was young (about 35), somewhat overweight, had muttonchop whiskers, and spoke very directly with a strong Liverpool accent. I mention these personal details only because it would seem that some people are adversely influenced by them. Acting as we were as proxies for Nature herself, we took care, therefore, not to allow ourselves to be distracted by extraneous issues. Seddon's appearance merely indicated to us that he could be an interesting person.[14]

In his opening remarks, Seddon told us that solvents had dominated the history of chemistry. In the earliest periods, water was *the* medium for the study of chemical reactions. More recently, a variety of new solvents have come into use—liquid ammonia, BrF_3, liquid halogens, and organic solvents. However, the majority have one common factor—they are all essentially molecular solvents with a covalent structure. He went on to tell us about a new class of solvents—ionic liquids.[15] They were transparent fluids at room temperature, stable over a wide range of temperatures (ranging from $-90°C$ to $+160°C$), and had relatively low viscosity, large electrochemical and spectroscopic windows, and no measurable vapor pressure up to about $100°C$. Hence, it was now possible, for the first time, to study inorganic, organometallic, and organic chemistry in a totally ionic environment without the problems of thermal degradation.

His proposed Venture Research contribution was therefore to study the chemistry of ions in an ionic environment. This contrasted with the study of molecules in a molecular environment that had dominated the history of chemistry to date. He also mentioned that he had previously sent the proposal to the Engineering and Physical Sciences Research Council (EPSRC). Their hypercritical response had been to give it a γ rating, *their lowest possible at the time*. I have never understood the research councils' need so precisely to categorize the research that they reject. In Seddon's case, a young scientist starting to make his independent way, the only

---

13 Seddon had been a junior member of one of the first Venture Research teams. With P. D. Calvert and A. J. McCaffrey (University of Sussex), he had studied "composites as active materials." Polymer matrices are ubiquitous in Nature. Materials such as mother-of-pearl, bone, and chitin are highly structured and designed to achieve specific purposes. The group wished to understand the ways that growth of such structures is controlled, and how they are designed.

14 One senior BP director was not impressed. When he saw us heading for the BP Visitors' Dining Room, which was indeed a very fine restaurant, he took me to one side and whispered: "You can't be taking him in there. He's not even wearing a tie!"

15 The liquid he started with was: $[emim]X-AlX_3$. $[emim]^+$ = 1-ethyl-3-methylimidazolium cation; X = Cl or Br.) He has subsequently used many others.

message it could convey was that his peers thought his ideas were utterly devoid of value. It was a contemptuous gesture, and put him in a position from which one would normally have found it very difficult to recover.

We thought differently. At the tactical level, Seddon's proposal seemed rich in possibilities. The ionic liquids he proposed to use were stable, pervaded by strong electric fields, and seemed to be virtually universal solvents for both ionic and molecular species. The behavior of neutral species in such liquids had hardly been studied. In addition, charged species remain intact, and have exceptionally low mobilities. Surprisingly, therefore, ionic liquids would provide relatively passive environments for the study of many important processes, particularly those involving transient or unstable species, and yield information not readily available using molecular solvents. In addition, they would avoid the solvation problem. Let us say that we wish to study the interaction between two charged molecular species A and B in a water solvent. As they move closer, the first interaction is not between A and B but between the water molecules that surround each of them. This extra complexity must be resolved before anything can be said about the primary interaction being studied. In an ionic solvent, however, there are no impediments; both A and B can "see" each other directly, and the interaction is uncontaminated.

We were also impressed by Seddon's strategy. He seemed to have discovered a major and virtually unexplored branch of chemistry—an intellectually green field—and seemed to know precisely how to go about exploring it. In short, we found the proposal irresistible, and we eventually persuaded Menter's board to fund it. However, we were soon reminded that the EPSRC's devastating response to Seddon's earlier submission could not be ignored, based as it was on the informed opinion of researchers deemed the best in the field. As is usual, they pass their judgments anonymously, of course, and so we could not know whether one of these experts might turn out to be the single peer whom we offered as our concession to peer review or one who might know a board member. We feared, therefore, that we might not have an easy ride, and we were right. The "peer" wrote a devastating assessment, and we could not get him to change his mind. But we went ahead anyway, and my submission to Menter's board included the dreadful review together with my own very positive recommendation. To make matters worse, the peer was known to a board member as someone who took pains to encourage young people! However, I was also a member, and so this time Seddon had a well-briefed scientific advocate who could argue his case. As might be expected, it was not straightforward, but to the board's credit, Seddon won the day.

Venture Research funding gave Seddon just the impetus he needed to lay the foundations of what turned out to be a new green technology. With a small team of PhD students and postdoctoral assistants, and interest from mainstream BP (Mike Green, Andy Fleet, and Martin Atkins), a new vista on chemistry opened up. Seddon did not rush into print, but with BP (and later Unilever and BNFL) filed a series of patents to protect the industrial sponsors of his work. These were over a broad range of applications, for polymerization of alkenes, alkylation of aromatics, synthesis of fragrances, preparation of fatty acids, and nuclear fuel reprocessing, illustrating something of the huge range of industrial opportunities in this area. In 1993, he moved to the chair of inorganic chemistry at the Queen's University of Belfast. When he finally published some of his work in a Russian journal in 1996 and in more accessible sources in 1997[16] and 1998, it triggered an explosion of interest. In 1997, 19 papers on ionic liquids were published; in 2006, over 2,028 appeared on this subject, and up to the time of writing (2007) the total has risen to over 8,000 in the past 10 years. It has now been estimated that there are well over a million possible ionic liquids, and over 1,018 derived ternary systems.

Seddon is now the most widely cited author in the field, and his organization QUILL (The Queen's University Ionic Liquid Laboratories, a university-academic consortium supported by 16 industries from four continents) is the most frequently cited research center. His work has been recognized by the president of the United States (awarded the 2005 Presidential Green Chemistry Challenge Award), and by the Queen (awarded the 2006 Queen's Anniversary Prize for Higher and Further Education for "Ionic Liquids: A Green Solution for Pollution"). Their work has also been formally recognized by the G8 nations (QUILL is the UK focus for the International Green Network), and the European Union (QUILL is a Marie-Curie Centre). It has attracted over £8 million (approximately $16 million) in support. Since BASF made the first public announcement of an industrial process based on ionic liquids (BASIL), another dozen have followed. In addition, there are now commercial suppliers of ionic liquids in most countries from the one-gram to the 50-ton scale.

---

16 When he wrote:
    The reactions we have observed represent the tip of an iceberg—all the indications are that room-temperature ionic liquids are the basis of a new industrial technology. They are truly designer solvents: either the cation or the anion can be changed, if not at will, then certainly with considerable ease, in order to optimize such phenomena as the relative solubility of the reactants and products, the reaction kinetics, the liquid range of the solvent, the cost of the solvent, the intrinsic catalytic behaviour of the media, and air-stability of the system. For the first time, it is possible to design a solvent to optimize a reaction (with control over both yield and selectivity), rather than to let the solvent dictate the course of the reaction. This, quite literally, revolutionizes the methodology of synthetic organic chemistry: it will never be the same again!

General References

1. M. J. Earle, J. M. S. S. Esperança, M. A. Gilea, et al., "The Distillation and Volatility of Ionic Liquids," *Nature* 439, no. 7078 (2006): 831–834.
2. M. Deetlefs, K. R. Seddon, and M. Shara, "Predicting Physical Properties of Ionic Liquids," *Physical Chemistry Chemical Physics* 8, no. 5 (2006): 642–649.
3. A. Arce, M. J. Earle, S. P. Katdare, et al., "Mutually Immiscible Ionic Liquids," *Chemical Communications* 24 (2006): 2548–2550.
4. M. Deetlefs, K. R. Seddon, and M. Shara, "Neoteric Optical Media for Refractive Index Determination of Gems and Minerals," *New Journal of Chemistry* 30, no. 3 (2006): 317–326.
5. C. M. Gordon, J. D. Holbrey, A. R. Kennedy, and K. R. Seddon, "Ionic Liquid Crystals: Hexafluorophosphate Salts," *Journal of Material Chemistry* 8, no. 12 (1998): 2627–2636.
6. C. J. Adams, M. J. Earle, G. Roberts, and K. R. Seddon, "Friedel-Crafts Reactions in Room Temperature Ionic Liquids," *Chemical Communications* no. 19 (1998): 2097–2098.
7. K. R. Seddon, "Ionic Liquids for Clean Technology," *Journal of Chemical Technology and Biotechnology* 68 (1997): 351–356.
8. K. R. Seddon, "Room Temperature Ionic Liquids: Neoteric Solvents for Clean Catalysis," *Kinetics and Catalysis* 37, no. 5 (1996): 743–748.

VR 19 and VR 22. Martyn Poliakoff (University of Nottingham) and Ken Seddon (Queen's University of Belfast)

The latter focus of their work has been the industrialization of green chemistry, and their research has been at what is conventionally regarded as the interfaces of chemistry, chemical engineering, biochemistry, and physics. But both passionately believe, with true Venture Research spirit, in the globalization of chemical advances. Their disparate solvents are now considered formally as neoteric solvents, and they have complementary rather than competitive applications.

No other Venture Researchers illustrate more clearly the benefits of striving to develop absolute selection criteria (in the absence of complete freedom as a right) instead of those based on myopic, conservative, and consensual peer review. This is especially necessary where young people are concerned. We can now see that our criteria equipped us with what might be seen as amazing foresight in supporting both of them, but of course foresight was not involved. We merely recognized their huge potential *to do something important*. Their impact on the science base and the chemical industry has indeed been substantial. There is no better way of mapping their recent careers than monitoring the impact of their publications (see **Figure 19**).

Citations, Poliakoff

Published Items in Each Year

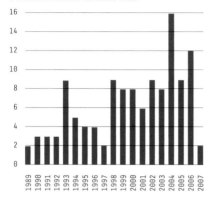

TS = (supercrit*)
AU = (poliakoff M)
DocType = All document types

Citations in Each Year

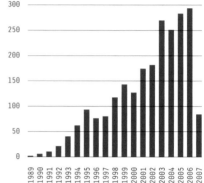

Language = All languages
Databases = SCI-EXPANDED, SSCI, A&HCI
Timespan = 1989–2007

Citations, Seddon

Published Items in Each Year

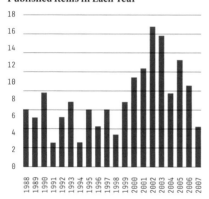

AU = (SEDDON KR)
DocType = All document types
Language = All languages

Citations in Each Year

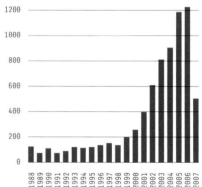

Databases = SCI-EXPANDED, SSCI, A&HCI
Timespan = 1988–2007

Results found: 263
Sum of the Times Cited *i*: 8,545
Average Citations per item *i*: 9
h-index *i*: 48

Figure 19: The evolution of publications and citations on Martyn Poliakoff's and Ken Seddon's Venture Research. (*Source: ISI Web of Science*, Copyright 2007, The Thompson Corporation.)

VR 23. Colin Self, Newcastle: "Biological Instability," 1990–1993

Nature depends on instability. In nucleic acids, the need for instability is well established—evolution depends on mutation and rearrangement of DNA. Without the ability to break down or to promote changes in the constituent elements of living systems, development and control would be so limited as to be useless. It was remarkable, therefore, that not only had instability and its consequences attracted little attention; experimentalists also tended to regard them as nuisances. They merely presented problems that had to be overcome.

To Colin Self, they represented substantial opportunity. Components of biological systems each have degrees of instability attached to their roles. As I explained in the chapter on the Damocles Zone, if a control element such as, say, the blood-clotting agent activated when we cut ourselves exhibits too little instability, it might exert its effect for too long, resulting, say, in our entire blood supply rapidly coagulating following even a minor accident. Conversely, if the element were too unstable, its effect could be transient and ineffective and blood loss could continue indefinitely. Thus, there is probably a spectrum of instabilities, some of which may currently be hidden from view because they are too transient. Plants, for example, do not appear to have active immune systems to protect them from infection, but they might be going undetected because they lie outside the spectrum of "visible" instabilities.

He was proposing to use the immune system as a model for understanding biological control in general. The immune system is highly regulated, but we knew almost nothing about its stability requirements. Exposure to a "foreign" antigenic substance must first lead to a recognition that it is indeed foreign, and that initial response might be followed by a commitment to produce antibodies highly specific to the antigen. Such distinct responses offer a rich spectrum of opportunities for study. To mention only a few:

- Do responses vary quantitatively or qualitatively with changes in antigen stability?
- How does antigen instability affect its effectiveness in causing an immune system to produce a specific antibody?
- How long must an antigen remain associated with the system for recognition to take place and a commitment to be made?
- To what extent is a potentially unstable antigen actually stabilized by association with receptors in the system?
- Is it possible to stabilize an antigen or speed up the commitment phase so that a highly unstable antigen causes a response?

As the immune system depends on the same principles of recognition as found in the complementary lock-key relationships between ligands and receptors that occur generally throughout Nature, answers to such questions could have wide biological significance and be truly transformative. Once again, we marveled at the funding agencies' myopia, and were delighted to give him our full and enthusiastic support.

A major problem in searching for specific antibodies against highly unstable materials is that they may be present only in exceedingly low concentrations—perhaps beyond the capability of the usual detection techniques. However, as serendipity would have it, we had supported Self a few years before in helping him develop, while still a medical student at Cambridge, a new form of immunoassay that went under the name *amplified* enzyme-linked immunoassay (AELIA). This phenomenally sensitive technique actually makes use of the natural amplification found in many biological control systems, and its new use brought Self's Venture Research contribution full circle. It had also led to a successful commercial development (by IQ Bio Ltd.) that made a great deal of money. If BP had had interest in that area, the profit from AELIA alone would have funded the entire cost of the Venture Research initiative up to 1985, the time of the initial development.

The most important outcome of his later work was the development of means of introducing *controlled instability* into biological systems. Medical scientists have long dreamed of controlling disease by "magic bullets" that will attack tumors, say, in a body, and do nothing else—that is, they would be free from all side effects. Self has made substantial progress toward achieving this end by developing his ideas on how instability works in general rather than by confronting specific disease directly, which is the conventional approach today. His work culminated in an experiment in which he first deactivated an antibody by associating it with a chemical that is also light-sensitive. The antibody was then injected into a body, and activated in the desired area only by illuminating it with low-intensity light (soft ultraviolet). Thus, the group showed that tumors in mice could be killed merely by shining a handheld light onto a particular location, and thereby became the first to demonstrate that antibodies can indeed be activated by light (Self and Thompson 1996). As antibodies can be raised against an enormous range of substances, this technology has very broad application in the development of ways to target and control a wide range of biological systems. When, for example, antibodies are directed against elements of the immune system (such as T cells), this means that immunity can be directed around the body by light. When used in conjunction with tumor-targeting antibodies, their inherent specificity can be dramatically increased. This enables a much greater differentiation between tumor cells and normal cells, reducing collateral damage and potentially serious side effects.

Self has shown that his unique approach to the control of biomedical systems is both viable and effective. One would have thought, therefore, that henceforth he would have little difficulty raising new funds. However, in today's risk-averse, bureaucratic environment, researchers starting outside the mainstream find that even when the efficacy of their new insight has been soundly demonstrated they must still spend some 90% of their time seeking new sponsorship. Transformative research initiatives must also address this important issue (which we also wanted to tackle, but BP forbade it) if their full value is to be realized.

VR 24. Gene Stanley, Boston; and José Teixeira, Laboratoire Léon Brillouin, CEA-CNRS, France: "Water in Confined Geometries," 1990–1993

When I first met Gene Stanley, we began by discussing liquid water. Its anomalous behavior is well known. Water expands, for example, at 4°C, when it is either heated or cooled, and its high dielectric constant means that it is a virtually universal solvent. These and its many other macroscopic properties are generally understood, but he went on to tell us that when water molecules are enclosed in spaces whose dimensions are comparable to the length of transient hydrogen-bonded networks, or when water is involved in reactions that take place over times comparable to the duration of a typical hydrogen bond, approximately 10^{-12} seconds, the behavior of water is hardly understood at all. Bearing in mind that water enclosed in living cells or in complex organic molecules is constrained in either or both these ways, the team seemed, therefore, to have identified a substantial shortfall in understanding. But why had these important problems not been addressed before? The chief obstacle was that the full range of systems in which water is tightly confined is diverse—clouds, rocks, as well as biological environments. Water had been well studied in each of the relevant disciplines—meteorology, geology, and biology—but researchers were interested in water only inasmuch as it impacted on the objectives of their discipline. Water per se was generally of little interest to them. Consequently, *coherent and impartial* studies on water structure and dynamics in confined geometries had been neglected simply because no one was interested in funding them. Despite its importance confined water, therefore, was a virtually unknown substance. The remarkable progress in understanding bulk water had come from focusing on the structure, dynamics, and statistics of the hydrogen-bonded network. However, these successful studies have assumed that bulk water exhibits long-range symmetry, which of course it does. But when the confining space is of the order of tens to hundreds of angstroms, this assumption is no longer valid, and the behavior of the hydrogen-bonded network is modified. The network will probably also hold the clue to understanding confined water and indeed might be a key to understanding how living cells accomplish their function. However, instead of considering water as a continuous medium, as had almost always been done in the past, the team wanted to consider the dynamics of the microscopic hydrogen-bonded network as an essential element of the cell.

Biological systems—cells, enzymes, and so on—are in constant vibrational or oscillatory motion at the molecular level, which the team thought would perturb the hydrogen-bonded network and could account for confined water's anomalous properties. They proposed to study the perturbations using two broad techniques—ultrasonic, and inelastic thermal-neutron scattering—chosen because they would probe different length scales of the network and so would yield complementary data. However, we were also impressed by the team's vision, particularly by their determination to understand biological systems from unusual perspectives. They suspected, for example, that randomness may play a key role in natural phenomena, a view that for us resonated with that of another budding Venture Research team—Colin Self (see VR 23)—and we were eager, therefore, to support it.

Taking full advantage of the freedom inherent in Venture Research, Stanley, Teixeira, and other subsequent collaborators launched a wide range of studies. They have now discovered a new liquid-liquid phase transition that would seem to explain the fundamental mechanisms underlying the remarkable behavior of liquid water. When water is rapidly cooled (at rates of around $10^6 K/s$), freezing is avoided, and at temperatures below about 140K becomes a noncrystalline solid; that is, a glass (Debenedetti and Stanley 2003). Glassy water is probably the most common form of water in the universe. It is observed as a frost on interstellar dust and constitutes the bulk of matter in comets. Furthermore, their discovery also applies to other liquids with a local tetrahedral structure, such as silicon and silica.

The observed transition (Mishima and Stanley 1998) takes place between a high-density liquid phase and a low-density liquid phase separated by a 30% jump-discontinuity in density. Stanley et al. have also developed a theoretical model for this transition. If the interaction energy between two molecules has a single well, the system will have a single critical point from which emanates a line of liquid-gas phase transitions. If, like water and tetrahedral liquids, the attractive well has two subwells, the outer of which is deeper and narrower, then for sufficiently low temperature, the one-phase liquid can condense into the narrower outer well, thereby giving rise to a new low-density liquid phase, and leaving behind the high-density liquid phase whose constituent molecules occupy predominantly the inner subwell.

As I mentioned earlier, it was hitherto believed that water's unusual behavior arose from the extensive network of hydrogen bonds. While this network is sufficient to explain some water properties, such as the relatively high melting and boiling points, it is not sufficient to explain others such as the apparent divergence of thermodynamic properties as supercooled water approaches 228K. Nor does it explain the properties of materials such as silica and silicon, which display anomalies similar to those of water but lack hydrogen bonds. The Stanley group's new mechanism explains these and other properties such as diffusivity and viscosity, which are also anomalous at low temperature. If liquid water is sufficiently cold, its diffusivity increases and its viscosity decreases on compression. In water, pressure disrupts the tetrahedral network, causing molecular mobility to increase. In contrast, for most other liquids compression leads to a progressive loss of fluidity as molecules are squeezed closer together.

The implications of the new liquid-liquid phase transition could be far-reaching. It not only explains many of water's properties but has also opened up the new field of *liquid polyamorphism*. Water's unique importance means that their work has implications for biology, astrophysics, materials science, and the technology of low-temperature preservation of biological molecules.

Stanley is a prolific author, and has pursued many interdisciplinary studies at the interfaces between physics and chemistry, physics and medicine, and even physics and finance. His papers have been cited 30,000 times, and his Hirsch index is 88. Only five other physicists have a higher index. However, Stanley wrote in 2007 to say that his Venture Research support played a vital role in his studies on water and other complex systems. Citing these studies, the Commission on Statistical Physics of the International Union of Pure and Applied Physics awarded him the Boltzmann Medal in 2004. Awarded every three years, the Medal is the highest award in statistical physics.

Gene Stanley was elected to membership in the US National Academy of Sciences in 2004.

VR 25. Harry Swinney, Texas at Austin; Werner Horsthemke, Southern Methodist University, Dallas; and Patrick De Kepper, Jean-Claude Roux, and Jacques Boissonade, Centre de Recherche Paul Pascal, Bordeaux: "Self-Organisation in Non-Linear Chemical Systems," 1985–1991

My first meeting with Harry Swinney was entirely serendipitous. I had traveled to the University of Texas at Austin to meet my old friend Peter Riley at the Physics Department, and Peter had introduced us. Swinney then told me of an experiment he had been planning for some time with a group of French chemists, but he had been unable to interest anyone in funding it. There are usually interesting possibilities in the offing when physicists and chemists want to work together, and this turned out to be no exception. Classically, when one wishes to study the interaction of two chemicals, say, one puts them in a test tube and stirs them thoroughly so that they are in intimate contact with each other.

Indeed, virtually all chemistry, academic or industrial, is done in what is effectively a zero-dimensional environment. That is what "well-stirred" implies. Bearing in mind that a scientist's main objective is to study the ways in which Nature does things and thereby to try to understand them, one should note that although natural phenomena are highly diverse, Nature would seem never to engage in well-stirred chemistry. No living system is well stirred, nor are Earth's rocks, oceans, or atmosphere or indeed the universe. There is structure everywhere we look.

Physicists and chemists had traditionally preferred to study systems in or near thermodynamic equilibrium where nonlinearities can be neglected or treated as small perturbations. In contrast, Nature (and industry) often proceeds by driving systems (by external gradients in temperature, concentration, velocity, etc.) far from equilibrium where nonlinearities play a crucial role. It is the nonlinear character of natural and technological systems that creates the remarkable similarities in the behavior of such diverse systems as fluid flows, lasers, and slime molds, and which in turn has given rise to the field of nonlinear dynamics. Indeed, the team first came together at one of the many conferences in this new field. Hitherto, most research on nonlinear phenomena had concerned *temporal* behavior, and the well-stirred reactor is sufficient for that. In biology and industry, however, diffusion processes and spatial properties can be crucial, but they cannot be studied in well-stirred reactors. Diffusion is, of course, a multidimensional process, but no one to date had devised a simple realizable prototype for processes involving diffusion.

The proposed Texas–Bordeaux collaboration had stemmed from contacts Horsthemke had made when he was a member of Prigogine's group in Brussels before he moved to the United States. The team had devised a new use for a very old apparatus—the circular Couette reactor, first devised in the late nineteenth century. In the Couette, reactions take place in the narrow annulus between concentric rotating cylinders into which reagents can be injected. When the differential rotation rates of the concentric cylinders exceed a certain value, toroidal Taylor vortices form, encircling the inner cylinder like coils of rope. With the continuous slow addition of chemicals either at one end of the annulus or separately to each Taylor vortex, each vortex becomes a separate reactor, coupled to its neighbors by diffusion. In another version of their apparatus, the two cylinders making up the Couette reactor would be rotated in opposite directions. In this latter case, there would be no Taylor vortices, but the turbulent diffusion rate in the axial direction could be controlled by varying the rotation rates of the two cylinders. The team proposed to study reactions of steadily increasing complexity as they developed this elegant technique. They would start with a few simple chemical reactions, then an autocatalytic reaction, and finally, address oscillating and chaotic reactions.

It was clear that neither group would dare to attempt such an ambitious program without the complementary expertise of the other. Furthermore, the proposed collaboration would not be physics, nor chemistry, nor chemical engineering, nor fluid mechanics but a combination of them all. As the team also straddled the Atlantic, funding from the usual sources became even less likely. Some preliminary work had been in progress for about a year, so they knew that the system worked. However, the full-scale collaboration could not take place without our support. We were delighted to oblige.

The group's long-range collaboration worked superbly well. Modern communications make it easier nowadays, but in addition, the collaboration was cemented by regular exchanges of postdocs between the two main centers. The Bordeaux team devised the idea of using the CIMA (chlorite-iodide-malonic acid) reaction as a model chemical system, and developed a one-dimensional chemical reactor that yielded the first Turing patterns. Shortly afterward, the Texas group observed two-dimensional Turing patterns in a study of the CIMA reaction in a higher-dimensional reactor, which BP patented. The group's goal from the outset had been to develop reactors to study laboratory-controlled and sustained chemical patterns, and indeed they went on to derive other geometries of reactors. Variants of their designs are now used in many laboratories worldwide. The key word here is "sustained." Transient patterns had been observed since the 1960s. Such patterns can be produced when chemicals are poured into a petri dish from which, say, a spiral might emerge and make a few rotations before the chemicals are consumed—that is, the system rapidly evolves toward equilibrium as the reaction runs down. However, virtually every interesting process used in industry or biology is systematically driven away from equilibrium by the continued arrival of new chemical feeds or food.

The group's Venture Research established a new field. As a measure of the impact of their work, the Web of Science relates that up to 2006, the four top most frequently cited papers from the Texas Venture Research, all published in *Nature* or *Science*, have been cited 831 times. These papers are listed below (under "General References"). Their next 11 most widely cited papers garnered another 921 citations. These citation numbers are huge for basic research in physics and chemistry. Medical research, for example, often has more numerous citations, but the number of participating researchers is much larger.

Harry Swinney was elected to membership in the US National Academy of Sciences in 1992.

General References
1. Z. Noszticzius, W. Horsthemke, W. D. McCormick, et al., "Sustained Chemical Waves in an Annular Gel Reactor: A Chemical Pinwheel," *Nature* 329, no. 6140 (1987): 619–620.
2. Q. Ouyang and H. L. Swinney, "Transition from a Uniform State to Hexagonal and Striped Turing Patterns," *Nature* 352, no. 6336 (1991): 610–612.
3. K. J. Lee, W. D. McCormick, Q. Ouyang, and H. L. Swinney, "Pattern Formation by Interacting Chemical Fronts," *Science* 261, no. 5118 (1993): 192–194.
4. K. J. Lee, W. D. McCormick, H. L. Swinney, and J. E. Pearson, "Experimental Observation of Self-Replicating Spots in a Reaction-Diffusion System," *Nature* 369, no. 6477 (1994): 215–218.
5. V. Petrov, Q. Ouyang, and H. L. Swinney, "Resonant Pattern Formation in Achemical System," *Nature* 388, no. 6643 (1997): 655–657.

VR 26. Robin Tucker, David Hartley, and Desmond Johnston, Lancaster: "Geometrodynamics," 1990–1993

It is rare today that one meets scientists yearning to tackle the truly great and profound problems. Such ambition is now generally discouraged by the need to show that progress to the order or timetable beloved by the funding agencies might indeed be possible. As such ambitious scientists are often theoreticians, claims that their work might lead to tangible outcomes are also difficult to sustain. Furthermore, one of the tragedies of modern science, as mentioned many times in this book, stems from the funding agencies' apparent belief that progress in science has now reached the stage where there are no longer any truly great unsolved problems. Thus, from this perspective, the grand plan has more or less been hammered out. Consequently, agencies do not need to provide the virtually unlimited freedom scientists once enjoyed, as the greater good would be better served by focusing research on perceived priorities among the myriad lesser problems that will always be with us.

As is well known, Einstein devoted many years attempting to produce a fully relativistic coherent theory of all the known forces of Nature—electromagnetic, nuclear, and gravitational. His dream was not to be realized. A few others have made similarly heroic attempts, but this supreme goal remains elusive. As a result, it is generally assumed that gravity, Nature's most pervasive force, can be neglected for all interactions on the atomic scale.

In the 1960s, John Wheeler attempted to realize Einstein's dream via a wholly geometric route. He was not successful, but Robin Tucker believed that the considerable progress made in mathematics since that time, to which he had contributed (on the theory of exterior differential systems and relativistic membranes), meant that a new attempt might be more successful. Thus, in response to the freedom that we could provide, he had drawn up an ambitious program that set out to derive a fully relativistic geometric description of natural phenomena. The breadth of his intellectual grasp, his infectious enthusiasm, and his audacity were impressive. As ever, I was mystified as to why the funding agencies would not be similarly impressed. The single peer we invited to provide the independent assessment of the proposed work (as required by our board) indeed confirmed that the very breadth of Tucker's proposal, ranging over problems on the micro- and macroscopic scales, would indeed constitute too rich a disciplinary mixture for the agencies to accommodate.

Not only did we agree to support him, but as the scope of his studies was so extensive and important, my intention at the time was that subject to our usual criteria on renewals, we would continue to support him for many years. We could not really expect him to succeed in a few years where Einstein had failed in almost as many decades. Sadly, however, the BP board meeting that approved their proposal turned out to be the last. From my perspective, therefore, our support soon came to an untimely end. However, the team's mathematical reformulation of inherently non-linear phenomena made important contributions to such fields as spinning strings, membrane theory, and the unification of the fundamental interactions, as summarized in the following list of publications:

1. D. H. Hartley and R. W. Tucker, "A Constructive Implementation of the Cartan-Kähler Theory of Exterior Differential Systems," *Journal of Symbolic Computation* 12, no. 6 (1991): 655–667.
2. M. Onder, T. Dereli, and R. W. Tucker, "Signature Transitions in Quantum Cosmology," *Classical and Quantum Gravity* 10, no. 8 (1993): 1425.
3. D. H. Hartley, R. W. Tucker, and P. Tuckey, "Equivalence of the Darboux and Gardner Methods for Integrating Hyberbolic Equations," *Duke Mathematical Journal* 77, no. 1 (1995): 167–192.
4. T. Dereli and R. W. Tucker, "A Broken Gauge Approach to Gravitational Mass and Charge," *Journal of High Energy Physics* 41, no. 3 (2002).
5. J. Gratus and R. W. Tucker, "A Quantum Geometric Description of a Q-Brane with Intrinsic Spin," *Nuclear Physics B* 88, nos. 1–3 (2000): 349–354.
6. R. W. Tucker, "Classical Field-Particle Dynamics in Space-Time Geometries," *Proceedings of the Royal Society of London: Mathematical, Physical, and Engineering Sciences* 460, no. 2 (2004): 2819–2844.

Robin Tucker's work might normally be regarded as of archetypal academic purity, and therefore of no practical relevance. However, Venture Research outcomes, in common with those from genuinely impartial pursuits of new knowledge in general, are unpredictable. We always invited Tucker to participate in our periodic workshops, therefore, set up to catalyze informal exchanges between Venture Researchers and the best of BP's industrial scientists and engineers. We took this course even though quantum gravity is hardly a hot topic in industrial circles. Most satisfyingly, however, these exchanges led to several long-term industrial contracts from BP and other companies for work in such areas as the stability of bridges under wind and rain excitations, friction forces on rotating drill strings, and fatigue damage to undersea marine risers due to the periodic emission of vortices. Eventually, their extensive understanding of nonlinear problems led to the foundation of the Industrial Mathematics and Gravity Group to promote further work in these areas. The work on one of their commissions is illustrated in **Figure 20**.

It is ironic, not to say astonishing, that even though the industrial value of this spin-off greatly exceeds what the team needs to continue on their Einsteinian quest, in today's short-term optimizing culture it is nevertheless not forthcoming.

Marine Riser Dynamics

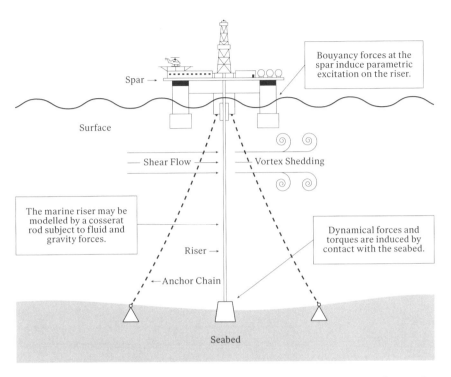

Figure 20: A cartoon illustrating the Industrial Mathematics and Gravity Group's work on marine riser dynamics. I hope that the reader may be both surprised and amused with this figure as it has evolved quite naturally from the purest academic research on quantum gravity. Perhaps Einstein, too, would be pleased. (Source: Reproduced by permission of Robin Tucker, David Hartley, and Desmond Johnston.)

Bibliography

Alvarez, Luis W. *Adventures of a Physicist*. New York: Basic Books, 1987.

American Association of University Professors. "Statement of Principles on Academic Freedom and Tenure, with 1970 Interpretive Comments." AAUP.org. Accessed May 20, 2020. https://www.aaup.org/report/1940-statement-principles-academic-freedom-and-tenure.

Amijee, F., E. J. Allan, R. N. Waterhouse, L. A. Glover, and A. M. Paton. "Non-Pathogenic Association of L-Form Bacteria (Pseudomonas Syringae pv. Phaseolicola) with Bean Plants (Phaseolus Vulgaris L.) and Its Potential for Biocontrol of Halo Blight." *Biocontrol Science and Technology* 2, no. 3 (1992): 203–214.

Banister, James A., Peter D. Lee, and Martyn Poliakoff. "Flow Reactors for Preparative Chemistry in Supercritical Fluid Solution: 'Solvent-Free' Synthesis and Isolation of $Cr(CO)_5(C_2H_4)$ and $(\eta^5-C_5H_5)Mn(CO)_2(\eta^2-H_2)$." *Organometallics* 14, no. 8 (1995): 3876–3885.

Barnett, Correlli. *The Audit of War*. London: Macmillan, 1986.

Berkshire, Frank H. In *Proceedings of the Conference on Statistics, Science and Public Policy*, edited by A. M. Herzberg, 136. Hailsham, UK: 2005.

Blackett, P. M. S. "The Rutherford Memorial Lecture, 1958." *Proceedings of the Royal Society of London Series A: Mathematical and Physical Sciences* 251, no. 1266 (1959): 293–305.

Bourdon, E. B. D. "Surface-Aligned Photochemistry." *Journal of Chemical Physics* 88, 6100 (1984).

Braben, Donald W. "What Makes an Engineer Tick." *New Scientist*, June 26, 1986.

Braben, Donald W. *To Be a Scientist*. Oxord: Oxford University Press, 1994.

Braben, Donald W. *Pioneering Research: A Risk Worth Taking*. Hoboken: Wiley, 2004.

Brenner, Sydney. "The Impact of Society on Science." *Science* 282, no. 5393 (1998): 1411–1412.

British Petroleum Company. *Our Industry Petroleum: A Handbook Dealing with the Organization and Functions of an Integrated International Oil Company with Particular Reference to the British Petroleum Company*. London: The British Petroleum Company Ltd., 1977.

Britt, Ronda. *Industrial Funding of Academic R&D Continues to Decline in FY 2004*. NSF 06-315. Arlington, VA: National Science Foundation, April 2006.

Brown, Harrison, James Bonner, and John Weir. *The Next Hundred Years*. London: The Scientific Book Club, 1957.

Buderi, Robert. *The Invention that Changed the World: The Story of Radar from War to Peace*. London: Abacus, 1996.

Bush, Vannevar. *Science—the Endless Frontier*. Washington, DC: National Science Foundation 40th Anniversary 1950–1990, 1990.

Caro, T. M., and M. D. Hauser. "Is There Teaching in Nonhuman Animals?" *Quarterly Review of Biology* 67, no. 2 (1992): 151–174.

Churchill, Winston S. "A Four Years' Plan, March 21, 1943." In *Onwards to Victory*, compiled by Charles Bade, 47–63. Boston: Little, Brown and Company, 1944.

Cicero, Marcus Tullius. *Cicero's Tusculan Disputations*. Translated by C. D. Yonge. New York: Harper & Brothers, 1877.

Cohen, I. Bernard. *Revolution in Science.* Cambridge, MA: Harvard University Press, 1985.

Committee on Higher Education. *Higher Education Report of the Committee appointed by the Prime Minister under the Chairmanship of Lord Robbins, 1961–63.* London: Her Majesty's Stationery Office, 1963

Cox, R. T., C. G. McIlwraith, and B. Kurrelmeyer. "Apparent Evidence of Polarization in a Beam of Beta-Rays." *Proceedings of the National Academy of Sciences USA* 14, no. 7 (1928): 544–549.

Cuatrecasas, Pedro. "Drug Discovery in Jeopardy." *Journal of Clinical Investigation* 116, no. 11 (2006): 2837–2842.

Daalder, Hans, and Edward Shils. *Universities, Politicians and Bureaucrats: Europe and the United States.* Cambridge: Cambridge University Press, 1982.

Dainton, Fred. *Doubts and Certainties: A Personal Memoir of the 20th Century.* Sheffield, UK: Sheffield Academic Press, 2001.

Darr, Jawwad A., and Martyn Poliakoff. "New Directions in Inorganic and Metal-Organic Coordination Chemistry in Supercritical Fluids." *Chemical Reviews* 99, no. 2 (1999): 495–541.

David, Peter. "Inside the Knowledge Factory," special report. *The Economist,* October 2, 1997.

Debenedetti, Pablo G., and H. Eugene Stanley. "Supercooled and Glassy Water." *Physics Today* 56, no. 6 (2003): 40–46.

Denison, Edward F. *Trends in American Economic Growth, 1929–1982.* Washington, DC: Brookings Institution, 1985.

Diamond, Jared. *Collapse: How Societies Choose to Fail or Succeed.* New York: Penguin Books, 2005.

Dicke, R. H., P. J. E. Peebles, P. G. Roll, and D. T. Wilkinson. "Cosmic Black-Body Radiation." *Astrophysics Journal* 142 (1965): 414–419.

Dijkstra, Edsger W. *Selected Writings on Computing: A Personal Perspective.* Berlin: SpringerVerlag, 1975.

Dobrin, Sergey, Krishnan R. Harikumar, Tingbin Lim, Lydie Leung, I. R. Mcnab, John C. Polanyi, Peter A. Sloan, et al. "Maskless Nanopatterning and Formation of Nanocrystals and Switches for Haloalkanes at $Si(111)-7 \times 7$." *Nanotechnology* 18, no. 4 (2007): 4012–4016.

Donofrio, Nicholas. "The Value of Innovation." *Ingenia* 22 (2005): 23-27.

The Economist. "Finance & Economics: Henry Hindsight." February 12, 2000.

The Economist. "The Global Housing Boom: In Come the Waves." June 16, 2005.

The Economist. "The Greening of General Electric: A Lean, Clean Electric Machine." December 10, 2005.

The Economist. "Higher Education: Remember Detroit," editorial. March 11, 2006.

The Economist. "The Rise and Fall of Corporate R&D: Out of the Dusty Labs." March 1, 2007.

Farman, J. C., B. G. Gardiner, and J. D. Shanklin. "Large Losses of Total Ozone in Antarctica Reveal Seasonal CIO_x/NO_x Interaction." *Nature* 315, no. 6016 (1985): 207–210.

Feynman, Richard, Robert B. Leighton, and Matthew Sands. *The Feynman Lecture Notes in Physics.* Reading, MA: Addison-Wesley, 1963.

Fleischmann, Martin, and Stanley Pons. "Electrochemically Induced Nuclear Fission of Deuterium." *Journal of Electroanalytical Chemistry* 261, no. 2 (1989): 301–308.

Fraga-Dubreuil, Joan, and Martyn Poliakoff. "Organic Reactions in High-Temperature and Supercritical Water." *Pure Applied Chemistry* 78, no 11 (2006): 1971–1982.

Franks, Nigel R., and Tom Richardson. "Teaching in Tandem-Running Ants." *Nature* 439, no. 7073 (2006): 153.

Freeman, Matthew. "Feedback Control of Intercellular Signalling in Development." *Nature* 408, no. 6810 (2000): 313–319.

Friis-Christensen, E., and K. Lassen. "Length of the Solar Cycle: An Indicator of Solar Activity Closely Associated with Climate Change." *Science* 254, no. 5023 (1991): 698–700.

Grampp, William D. "What Did Smith Mean by the Invisible Hand?" *Journal of Political Economy* 108, no. 3 (2000): 441–64.

Green, John Richard. *History of the English People*. London: Macmillan, 1909.

Hahn, Roger. *The Anatomy of a Scientific Institution: The Paris Academy of Sciences, 1666–1803*. Los Angeles: University of California Press, 1971.

Howdle, Steven M., Michael A. Healy, and Martyn Poliakoff. "Organometallic Chemistry in Supercritical Fluids—the Generation and Detection of Dinitrogen and Nonclassical Dihydrogen Complexes of Group-6, 7, and 8 Transition Metals at Room Temperature." *Journal of the American Chemical Society* 112, no. 12 (1990): 4804–4813.

Hurley, John. *Organisation and Scientific Discovery*. Chichester, UK: Wiley, 1997.

Hutchison, Terence. "Adam Smith and The Wealth of Nations." *Journal of Law and Economics* 19, no. 3. (1976): 507–528.

Kuhn, Thomas S. *The Structure of Scientific Revolutions 2nd ed*. Chicago: University of Chicago Press, 1970.

Kunze, E., J. F. Dower, I. Beveridge, R. Dewey, and K. P. Bartlett. "Observations of Biologically-Generated Turbulence in a Coastal Inlet." *Science* 313, no. 5794 (2006): 1768–1770.

Lane, Nick. *Power, Sex, Suicide: Mitochondria and the Meaning of Life*. Oxford: Oxford University Press, 2005.

Lederman, Leon. *Science: The End of the Frontier*. Washington, DC: American Association for the Advancement of Science, 1991.

Lee, T. D., and C. N. Yang. "Question of Parity Conservation in Weak Interactions." *Physical Review* 104, no. 1 (1956): 254–258.

Licence, Peter, Je Kie, Maia Sokolova, Stephen K. Ross, and Martyn Poliakoff. "Chemical Reactions in Supercritical Carbon Dioxide: From Laboratory to Commercial Plant." *Green Chemistry* 5, no. 2 (2003): 99–104.

Maddox, John. *What Remains to Be Discovered: Mapping the Secrets of the Universe, the Origins of Life, and the Future of the Human Race*. London: Macmillan, 1998.

"Magna Charta Universitatum." Observatory Magna Charta Universitatum. Bologna: September 18, 1988. Via http://www.magna-charta.org/resources/files/the-magna-charta/english.

McCarthy, J. "Towards a Mathematical Science of Computation." In *Computer Programming and Formal Systems*, edited by P. Braffort and D. Hirschberg. Amsterdam: North-Holland, 1963.

Mishima, Osamu, and H. Eugene Stanley. "The Relationship between Liquid, Supercooled and Glassy Water." *Nature* 396, no. 6709 (1998): 329–335.

Morison, Samuel Eliot. *Three Centuries of Harvard, 1636–1936*. Cambridge, MA: Belknap Press, 1969.

Morrissey, Susan R. "Elias A. Zerhouni." *Chemical & Engineering News* 84, no. 27 (2006): 12–17.

Mowery, David C., and Nathan Rosenberg. *Technology and the Pursuit of Economic Growth*. Cambridge, UK: Cambridge University Press, 1989.

National Institutes of Health. "NIH Director's Pioneer Award Program." 2008. Announcement, RFA-RM-08-013. https://grants.nih.gov/grants/guide/rfa-files/RFA-RM-08-013.html.

National Science Board. *2020 VISION for the National Science Foundation.* December 28, 2005. National Science Foundation, NSB-05-142. https://www.nsf.gov/attachments/106772/public/National_Science_Board_2020_Vision_nsb05142.pdf.

National Science Foundation. "Guide to Programs." NSF.gov. n.d. https://www.nsf.gov/od/lpa/news/publicat/nsf0203/cross/ocpa.html.

"New Radio Waves Traced to Centre of the Milky Way." *The New York Times.* May 5, 1933.

OECD Main Science and Technology Indicators (MSTI) Database. Accessed 2007. https://www.oecd.org/sti/msti.htm.

Olson, Åke, and Jan Stenlid. "Plant Pathogens: Mitochondrial Control of Fungal Hybrid Virulence." *Nature* 411, no. 6836 (2001): 438.

Owen, Tom. "The University Grants Committee." *Oxford Review of Education* 6, no. 3 (1980): 255–278.

Parkhouse, J. G. "Structuring: A Process of Material Dilution." *In Proceedings of the Third International Conference on Space Structures,* edited by H. Nooshin. London: Elsevier Applied Science, 1984.

Parkhouse, J. G. "Damage Accumulation in Structures." *Reliability Engineering* 17, no. 2 (1987): 97–109.

Parkhouse, J. G. "The Influence of Prestress on Composite Performance." In *Applied Solid Mechanics* Vol. 3, edited by I. M. Allison and C. Ruiz. London: Elsevier Applied Science, 1989.

Parkhouse, J. G., and H. R. Sepangi. "Macromolecules." In *Proceedings of the Seminar: Building the Future, Brighton, April 1993,* Fig. 10.207. London: E. & F. N. Spon, 1994.

Parkhouse, J. G., H. R. Sepangi, and W. E. Williams. "Structural Simplicity Through a Lenticular Pattern." *International Journal of Mechanical Sciences* 34, no. 12 (1992): 957–970.

Pendry, J. B. "Symmetry and Transport of Waves in One-Dimensional Disordered Systems." *Advances in Physics* 43, no. 4 (1994): 461–542.

Pendry, John Brian, A. MacKinnon, and P. J. Roberts. "Universality Classes and Fluctuations in Disordered Systems." *Proceedings of the Royal Society: Mathematical and Physical Sciences* 437, no. 1899 (1992): 67–83.

Penzias, A. A., and R. W. Wilson. "A Measurement of Excess Antenna Temperature at 4080 Mc/s." *Astrophysical Journal* 142 (1965): 419–421.

Planck, Max. *Where Is Science Going?* Translated by James Murphy. Woodbridge, CT: Ox Bow Press, 1933.

Polanyi, Michael. *Knowing and Being.* London: Routledge & Kegan Paul, 1969.

Poliakoff, Martyn, Steven M. Howdle, and Sergei Kazarian. "Vibrational Spectroscopy in Supercritical Fluids: From Analysis and Hydrogen-Bonding to Polymers and Synthesis." *Angewandte Chemie English Edition* 34, no. 12 (1995): 1275–1295.

Poliakoff, Martyn, J. Michael Fitzpatrick, Trevor R. Farren, and Paul T. Anastas. "Green Chemistry: Science and Politics of Change." *Science* 297, no. 5582 (2002): 807–810.

Robinson, Elva J. H., Duncan E. Jackson, Mike Holcombe, and Francis L. W. Ratnieks. "Insect Communication: 'No Entry' Signal in Ant Foraging." *Nature* 438, no. 7067 (2005): 442.

Ronayne, J. *Science in Government.* London: Edward Arnold, 1984.

Ross, Ian K. "Mitochondria, Sex, and Mortality." *Annals of the New York Academy of Sciences* 1019, no. 1 (2004): 581–584.

Saward, D. *Bernard Lovell: A Biography*. London: Robert Hale, 1984.

Schlesinger, William H. Editorial, "Carbon Trading." *Science* 314, no. 5803 (2006): 1217.

Self, Colin H., and Stephen Thompson. "Light Activatable Antibodies: Models for Remotely Activatable Proteins." *Nature Medicine 2*, no. 7 (1996): 817–820.

Smith, Adam. *The Theory of Moral Sentiments*. London: A. Milar, and A. Kincaid and J. Bell, 1759.

Smith, Adam. *An Inquiry into the Nature and Causes of the Wealth of Nations*. London: W. Strahan and T. Cadell, 1776.

Smolin, Lee. *The Trouble with Physics: The Rise of String Theory, the Fall of a Science, and What Comes Next*. London: Allen Lane, 2007.

Solow, Robert. "Growth Theory and After." In *Nobel Lectures: Economic Sciences, 1981–1990*, edited by Karl-Göran Mäler, 199–213. Singapore: World Scientific Publishing, 1992.

Stephens, Michael Dawson, and Gordon Wynne Roderick, eds. *Universities for a Changing World: The Role of the University in the Late Twentieth Century*. Devon: David & Charles, 1975.

SQW, Ltd. *Impact of Quality-Related (QR) Funding for Research in English Higher Education Institutions*. January, 2007. https://www.sqw.co.uk/files/9513/8712/1370/128.pdf.

Stickland, Tim R., Nicholas F. Britton, and Nigel R. Franks. "Models of Information Flow in Ant Foraging: The Benefits of Both Attractive and Repulsive Signals." In *Information Processing in Social Insects*, edited by Claire Detrain, Jean-Louis Deneubourg, and Jacques M. Pasteels, 83–100. Basel: Birkhauser-Verlag, 1999.

Strangway, D. W. "University Reform is Global." In *Proceedings of the Conference on Statistics, Science and Public Policy, Held at Herstmonceux Castle, Hailsham, UK, April 2005*, edited by A. M. Herzberg, 97–100. Kingston, Ontario: Queen's University, 2006.

Thomas, J. M., "What is Happening to Our Universities?" Lecture presented at the Honourable Society of Cymmrodorion to mark its 250th anniversary, The Royal Society, London, 2001.

Thorsteinsson, Thorstein, David L. Cooper, Joseph Gerratt, and Mario Raimondi. "A New Approach in Valence Bond Calculations: CASVB." In *Quantum Systems in Chemistry and Physics: Trends in Methods and Applications*, edited by Roy McWeeny, Jean Maruani, Yves G. Smeyers, and Stephen Wilson, 67–85. Dordrecht, Netherlands: Kluwer Academic Publishers, 1997.

Tofts, Chris, and Nigel R. Franks. "Doing the Right Thing: Ants, Honeybees, and Naked Mole Rats." *Trends in Ecology & Evolution* 7, no. 10 (October 1992): 346–349.

Townes, Charles H. *How the Laser Happened: Adventures of a Scientist*. Oxford: Oxford University Press, 1999.

UK Council for Scientific Policy. *Report on Science Policy*, presented to the secretary of state for education and science May 1966.

UK University Grants Committee. *University Development 1957 to 1962*. White paper, 1962. Cmnd 2267, para. 627.

US Congress. House of Representatives, Committee on Science. *Unlocking Our Future: Toward a New National Science Policy*. 105th Cong., September 1998. Com. Pr. 105-B. https://www.aaas.org/sites/default/files/s3fs-public/GPO-CPRT-105hprt105-b.pdf.

Walker, R., C. M. J. Ferguson, N. A. Booth, and E. J. Allan. "The Symbiosis of Bacillus Subtilis L-Forms with Chinese Cabbage Seedlings Inhibits Conidial Germination of Botrytis Cinerea." *Letters in Applied Microbiology* 34, no. 1 (2002): 42–45.

Waugh, W. L. Jr. "Issues in University Governance: More 'Professional' and Less 'Academic.'" *Annals of the American Academy of Political and Social Science* 585 (January 2003): 84–96.

Widom, A., and T. D. Clark. "Quantum Electrodynamic Uncertainty Relations for Magnetic Flux Measurements in Circuits." *Journal of Physics A: Mathematical and General* 15, no. 11 (November 1982): 3617–3621.

Williams, Garnett P. *Chaos Theory Tamed.* London: Taylor & Francis, 1997.

Wilson, Robert W. "The Cosmic Microwave Background Radiation." In *Nobel Lectures: Physics, 1971–1980,* edited by Stig Lundqvist, 463–483. Singapore: World Scientific Publishing, 1992.

Wood, Peter. *Poverty and the Workhouse in Victorian Britain.* Gloucester: Alan Sutton, 1991.

Woodall, Pam. "The New Titans: A Survey of the World Economy." *The Economist,* September 16, 2006.

Wu, C. S., E. Ambler, R. W. Hayward, D. D. Hoppes, and R. P. Hudson. "Experimental Test of Parity Conservation in Beta Decay." *Physics Review* 105, no. 4 (1957): 1413–1415.

Young, Davis A. *Mind over Magma: The Story of Igneous Petrology.* Princeton, NJ: Princeton University Press, 2004.

Ziman, J. M. "The Bernal Lecture, 1983: The Collectivization of Science." *Proceedings of the Royal Society of London: Mathematical and Physical Sciences* A 389, no. 1796 (1983): 213.

Zuckerman, Solly. *Scientists and War.* London: The Scientific Book Club, 1996.

Index

Note: references to *footnotes* are indicated as 5n, 100n; *figures* are indicated as 5f, 100f; *tables* are indicated as 5t, 100t; and *posters* as 5p, 100p.

Notes from the Art Department

Creative Director: Tyler Thompson
Typesetting & Design: Kevin Wong
Printing: Hemlock Printers Ltd.
Bindery: Roswell Bookbinding

Composed in Ivar from Letters From Sweden;
and Record Gothic Mono by A2-Type

Printed on 60# Lynx

Bound in Pearl Linen

Inks:
■ Pantone 186U
■ Black

Stripe
Press